THE LIBRARY OF HOLOCAUST TESTIMONIES

Fire without Smoke

The Memoirs of a Polish Partisan

The Library of Holocaust Testimonies

Editors: Antony Polonsky, Sir Martin Gilbert CBE, Aubrey Newman, Raphael F. Scharf, Ben Helfgott MBE

Under the auspices of the Yad Vashem Committee of the Board of Deputies of British Jews and the Centre for Holocaust Studies, University of Leicester

My Lost World by Sara Rosen
From Dachau to Dunkirk by Fred Pelican
Breathe Deeply, My Son by Henry Wermuth
My Private War by Jacob Gerstenfeld-Maltiel
A Cat Called Adolf by Trude Levi
An End to Childhood by Miriam Akavia
A Child Alone by Martha Blend
The Children Accuse by Maria Hochberg-Marianska and Noe Gruss
I Light a Candle by Gena Turgel
My Heart in a Suitcase by Anne L. Fox
Memoirs from Occupied Warsaw, 1942–1945
by Helena Szereszewska
Have You Seen My Little Sister?
by Janina Fischler-Martinho
Surviving the Nazis, Exile and Siberia by Edith Sekules
Out of the Ghetto by Jack Klajman with Ed Klajman
From Thessaloniki to Auschwitz and Back
by Erika Myriam Kounio Amariglio
Translated by Theresa Sundt
I Was No. 20832 at Auschwitz by Eva Tichauer
Translated by Colette Lévy and Nicki Rensten
My Child is Back! by Ursula Pawel
Wartime Experiences in Lithuania by Rivka Lozansky Bogomolnaya
Translated by Miriam Beckerman
Who Are You, Mr Grymek? by Natan Gross
Translated by William Brand
A Life Sentence of Memories by Issy Hahn, Foreword by Theo Richmond
An Englishman in Auschwitz by Leon Greenman
For Love of Life by Leah Iglinsky-Goodman
No Place to Run: The Story of David Gilbert by Tim Shortridge and Michael D. Frounfelter
A Little House on Mount Carmel by Alexandre Blumstein
From Germany to England Via the Kindertransports by Peter Prager
By a Twist of History: The Three Lives of a Polish Jew by Mietek Sieradzki
The Jews of Poznań by Zbigniew Pakula
Lessons in Fear by Henryk Vogler
To Forgive … But Not Forget by Maja Abramowitch

Fire without Smoke

The Memoirs of a Polish Partisan

FLORIAN MAYEVSKI

with Spencer Bright

VALLENTINE MITCHELL
LONDON • PORTLAND, OR

First published in 2003 in Great Britain by
VALLENTINE MITCHELL
Crown House, 47 Chase Side
London N14 5BP

and in the United States of America by
VALLENTINE MITCHELL
c/o ISBS, 5824 N. E. Hassalo Street
Portland, Oregon 97213-3644

Website: www.vmbooks.com

British Library Cataloguing in Publication Data

ISBN 0-85303-461-3 (paper)
ISSN 1363-3759

Library of Congress Cataloging-in-Publication Data

A catalog record for this book is available from the Library of Congress

Typeset in 11/12.25pt Zapf Calligraphy by
Frank Cass Publishers Ltd, London
Printed by MPG Books Ltd, Bodmin, Cornwall

I dedicate this book to my cherished mother Rachel, sister Sara and brother Szulim, who perished under the Nazi occupation.

And to those who helped me to survive at the risk of their own lives: the inhabitants of Siucice and neighbouring villages.

Contents

Acknowledgements

For her love and support, my wife Sylvia.

Spencer Bright for helping me realise a long-held ambition.

Lilian King, for helping with translation. My daughter Kamila for her support and helping organize the Polish translation. The Holocaust Survivor's Centre Creative Writing Group and its head, Andrew Herskovitz. And for their contributions and advice: Tony Goodman; Stephen and Kate Annett; Professor Dr Feliks Tych, Institute of Jewish History, Warsaw; and Ruth Chern.

Spencer Bright would like to thank: Chrissy Iley, Susan Hill, Marianne Velmanns, Julian Alexander, Kevin O'Brien, and Rod Digges.

Spencer Bright is a journalist and author of *Take It Like a Man: The Autobiography of Boy George*; *Peter Gabriel: An Authorized Biography*; *You're Barred, You Bastards! The Memoirs of a Soho Publican*; and *Essential Elton*. Spencer Bright's father, Michael Brzegowski, was born in Lodz, leaving at the start of the Second World War. He was the sole survivor of his family.

List of Illustrations

Between pages 76 and 77

List of Maps

The Library of Holocaust Testimonies

It is greatly to the credit of Frank Cass that this series of survivors' testimonies is being published in Britain. The need for such a series has been long apparent, where many survivors made their homes.

Since the end of the war in 1945 the terrible events of the Nazi destruction of European Jewry have cast a pall over our time. Six million Jews were murdered within a short period; the few survivors have had to carry in their memories whatever remains of the knowledge of Jewish life in more than a dozen countries, in several thousand towns, in tens of thousands of villages, and in innumerable families. The precious gift of recollection has been the sole memorial for millions of people whose lives were suddenly and brutally cut off.

For many years, individual survivors have published their testimonies. But many more have been reluctant to do so, often because they could not believe that they would find a publisher for their efforts.

In my own work over the past two decades I have been approached by many survivors who had set down their memories in writing, but who did not know how to have them published. I also realized, as I read many dozens of such accounts, how important each account was, in its own way, in recounting aspects of the story that had not been told before, and adding to our understanding of the wide range of human suffering, struggle and aspiration.

With so many people and so many places involved, including many hundreds of camps, it was inevitable that the historians and students of the Holocaust should find it difficult at times to grasp the scale and range of events. The publication of memoirs is therefore an indispensable part of the extension of knowledge, and of public awareness of the crimes that had been committed against a whole people.

Sir Martin Gilbert
Merton College, Oxford

Preface

For over 50 years what I went through in Poland during the War has been ever present in my memory. Few people outside my family had heard of my experiences. The villagers of Siucice, Zawada and Chorzew knew about the Jewish boy who fought the Nazis. The villagers of Blogie and thereabouts knew I was a partisan leader, but they didn't know I was Jewish. For many years I have wanted the wider world to know my story.

I first wrote down my experiences in Polish in the 1960s when I still lived in Poland. I didn't show my account to anybody because I was afraid the then oppressive Communist government would disapprove of me having fought on the side of a non-communist group. I had become assimilated into Polish society and few people even knew that I was Jewish.

I felt betrayed when a new wave of anti-Semitism swept Poland in 1968 stoked up by the Communist Party. Hounded for being a Jew once more I no longer felt welcome in my homeland and resolved, in my forties, to go into exile.

It was not until I settled in England that I again thought about committing my memory to paper. But I had a new life to build and it was not until my retirement in 1987 that I thought 'now is my chance to tell my story before I die'.

At first I thought I would get someone to translate my account into English. But it became clear that the only way to do the task properly, despite my imperfect English, was to write in my adopted tongue. It took over a decade to get here.

I am not writing this to heal my wounds because my wounds will never heal. If anything, writing this book has brought all the pain back.

As I was nearing completion of the book I began having violent nightmares. I was fighting Nazis, running away, in turmoil over how to help my family. I would thrash out hitting my poor wife Sylvia and disturbing her sleep. Several times I fell out of bed. Sylvia was frightened by my shouting and had to wake me because I was unaware of what I was doing. In the end I had to seek help from a psychologist. I have found it does not get easier with time; you just have to learn how to live with it.

I wanted my surviving family and future generations to understand that not all Jews accepted their fate without resistance to the Nazis. Here was one Jew who led a group of men in armed resistance and survived. I survived through strength of will, luck, and the help of many Poles to whom I will be forever grateful.

Florian Mayevski
London, March 2001

1. Poland's borders from September 1939 to June 1941. Part of occupied Poland was annexed to Greater Germany while the rest of the country was designated as the General Government. The USSR administered an area east of the River Bug.

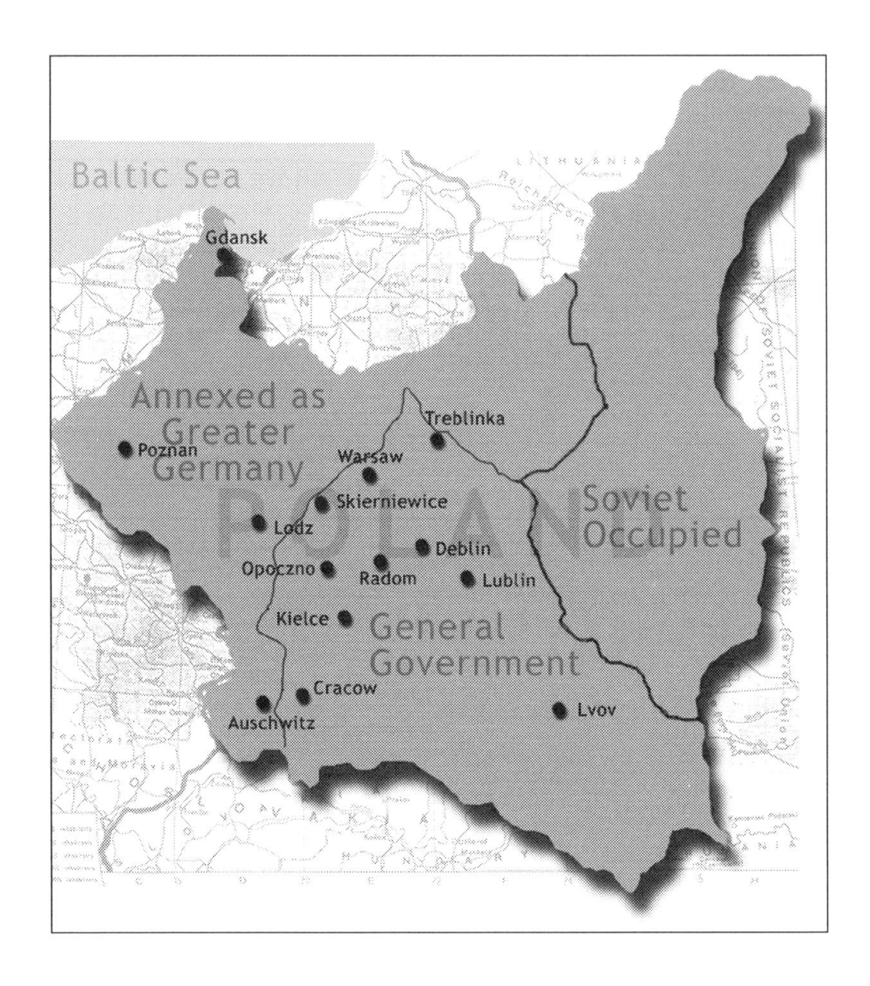

2. The region of central Poland where I sought refuge and fought during the Second World War. I was born in Sulejow and fled with the family to Siucice. Later, I built my bunker in the forest outside Blogie and based my camp there on becoming a hermit and a partisan.

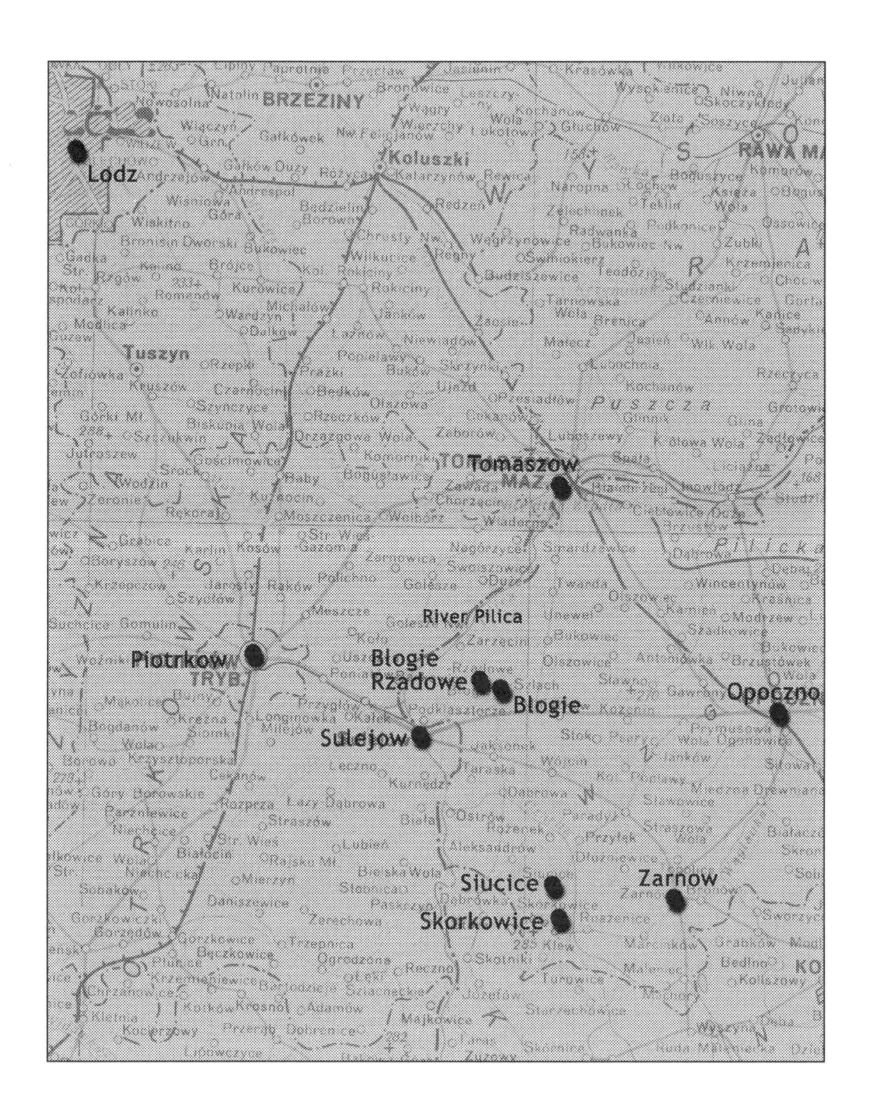

Introduction

This fascinating memoir is an unusual and important contribution to our knowledge of the fate of Polish Jews and to the history of Poland under Nazi occupation. Moshe Aaron Lajbcygier, who during the War took the Polish name Florian Majewski – anglicized after his emigration to the UK in 1973 to Mayevski – was better placed than most Polish Jews to survive the Nazi occupation in which more than 90 per cent of the country's Jewish population lost their lives.

He was born in 1922 in the small central Polish town of Sulejow, which had a population at the outbreak of the Second World War of around 6,000, half of whom were Jews. There were two schools in the town and Moshe attended the one in which most of the pupils were Catholic. As a result, he had many Polish friends (his best friend, Mirek Kuczkowski, was a Pole), spoke Polish like a native, and was often taken for a Catholic. He left school at 14 and served a two-year apprenticeship as a carpenter. He also learnt the family trade of baking from his uncles.

His fluency in Polish and his skills as a carpenter and baker stood him in good stead when the Nazis occupied the western parts of Poland in autumn 1939. In March 1940, he moved with his mother and siblings to the village of Siucice, about 25 kilometres from his native town. There they found a farmer who was willing to provide them with lodgings and he made himself useful in the village as a carpernter, while his mother sold haberdashery and other household items. The whole family was able to shelter in this rural backwater until 1941, when the Nazis ordered all Jews in the villages to assemble in the neighbouring town of Zarnow. Moshe was able to escape; the rest of his family perished.

It was now that his pre-war contacts proved invaluable. Moshe had been friendly with Marian Kowalski, a man in his early forties, who lived in another village, Blogie. When Moshe escaped from Zarnow, he first sheltered in Siucice where he continued to work as a carpenter and then, when the situation became unsafe, went to Marian for advice. His friend advised him to build himself a bunker (and then a second) in the nearby forests, which he did, and where he lived as a hermit until early 1943, using his skills in carpentry to construct a hideout, and receiving assistance from Marian.

In the spring of 1943, Marian, who had been active in the Home Army (Armia Krajowa – AK – the underground movement loyal to the Polish Government in London), observed Florian's skills in living in the forest and recruited him for the AK, counselling him to change his name (to Florian Majewski).

This is the most unusual part of Moshe/Florian's memoirs. Poland, during the Second World War, was the scene of bitter clashes, not only between partisan forces and the Nazi occupiers, but between those forces loyal to the London government and the authorities it established in occupied Poland and those attempting to establish a communist regime in that country (disguised as a Popular Front). In early 1942, as relations with the Polish Government in London – which had been re-established after the Nazi invasion of the Soviet Union in June 1941 worsened – Stalin gave permission to Polish communists in the USSR to reform the Polish Communist Party (dissolved on Stalin's orders in 1938). The new party, called the Polish Workers' Party (Polska Partia Robotnicaz – PPR) was intended both as a means of putting pressure on the London government to be more amenable, and also as a nucleus of a pro-Soviet Government in Poland, should an accommodation with the government in London prove impossible. The PPR pursued a 'popular front' policy in occupied Poland, but soon found itself in bitter conflict with the underground forces linked with the London government. It also established an underground military formation, first called the People's Guard (Gwardia Ludowa) and then the People's Army (Armia Ludowa).

These developments are the background for the clashes

between the two underground groups described in this memoir. The communist-controlled forces, which were quite weak, advocated an immediate confrontation with the Nazi occupier, both to take the pressure off the Soviet Union and in order to radicalise the situation in Poland by courting savage German reprisals. The Home Army wanted to avoid a major confrontation, partly to spare the civilian population, but above all because it wanted to harbour its strength until the decisive moment when the German power was collapsing. Its aim was to use this moment to take power in Poland and then confront the Soviets with the alternatives of negotiating with the London government and its forces in Poland, or of crushing them in the eyes of the world. This was a risky strategy and, as is well known, it failed disastrously and was followed by the sovietization of Poland. It was dictated by the desperate strategic position of the country and by the realisation on the part of the London Poles that they had very little chance of returning to Poland.

The attitude of the forces that made up the Polish Underground State to the Jews, and to the anti-Jewish genocide carried out by the Nazis, has been a matter of controversy. The tragic fate of the Jews did arouse considerable sympathy in the central bodies of the Underground. This was clearly expressed in the principal paper of the Underground, *Biuletyn informacyjny*, a weekly that appeared throughout the War and which reflected the views of the co-ordinating body of the civilian Underground, the Government Delegation, and (at its head) the Government Delegate. At the same time, within the Home Army, there was a determination to avoid premature military action and to conserve its strength (and weapons) for the crucial confrontation that would determine the fate of Poland. In addition, as the Commander of the Home Army, General Grot-Rowecki made clear, in an order of 10 November 1942, that they did not regard the Jews as 'part of our nation', and action was not to be taken to defend them if it endangered other AK objectives. Certainly the Home Army was not willing to absorb the Jewish partisan groups formed in the forests by fugitives from the ghettos, regarding them as unreliable and potentially communist in sympathy. There was one exception to this. In

Volhynia, which was wracked by a brutal ethnic conflict between Poles and Ukrainians, the AK was eager to co-operate with Jewish partisans to defend Polish villages. However, it was not, by and large, willing to accept Jews as individuals, though here too there were exceptions, such as the Propaganda and Information Bureau of the High Command. It should be mentioned, too, that the Home Army, like the civilian Underground, was made up of adherents of different political orientations, some of them sympathetic and others hostile to the Jews. The AK was not sympathetic to the plight of individual Jewish fugitives, seeing them as security risks likely to endanger its own position. Local commanders and the High Command often referred to these people (and also to communist partisans) as 'bandits'; an echo of the language used by the Nazis themselves.

The small military formations linked with the various fascist groups, the National Armed Forces (Narodowe Sily Zbrojne –NSZ) and the Rampart Group (Grupa Szażca) were openly hostile to the Jews, and were frequently guilty of murders both of Jewish partisans and of Jews hiding in villages. This situation continued even when the NSZ became more closely linked with the Home Army towards the end of the War.

The People's Guard and its successor, the People's Army, were much more willing to absorb Jews, both because in their isolation they needed any support they could obtain, and because their ideology stressed the importance of transcending national divisions. This was of course a mixed blessing, because the more Jews supported these groups, the more they seemed to confirm the belief in the Home Army (and elsewhere in Poland) that they were essentially siding with the communists.

Given this complex situation, the presence of a Jew in a Home Army formation in Central Poland was an unusual phenomenon and reveals some interesting details of a still largely unresearched topic. Marian swore in his small group. At this stage, they were desperate for recruits and even accepted a number of escaped Soviet prisoners-of-war. Partisan warfare as described by the author appears as a rather messy business. The unit was intended to interrupt

German communications, but for the most part occupied itself in acquiring weapons (often by force) and dealing with German collaborators, who were often executed.

Matters became more complicated in the spring of 1944. By this stage, more partisan groups began to operate in the hilly area south of Opoczno, some of them quite large. The Home Army was a very decentralised body and some of the constituent groups were very right wing in character. Florian first scented danger when an officer in one of these groups heard him talk in Russian to a member of his group:

'You are still keeping them alive?'

I said, 'We are all fighting the same enemy.'

The officer laughed and said, 'We finished them off a long time ago.'

Questions were also now raised about Florian's 'questionable' (i.e. Jewish) origins. After consulting with Marian he decided, together with the Russians in his group, to join a People's Army unit operating in the area. This AL group, about 50-strong, was largely Polish, although it also included about ten Soviet parachutists and two German veterans of the Spanish Civil War: it was commanded by a Soviet officer, Major Gromov, who took radio orders from Kiev. The group undertook a number of actions against German forces, and blew up army railway transports.

When the War ended, Florian searched unsuccessfully for members of his family in Sulejow and Lodz. He entered the Polish Army with the rank of sergeant and eventually attained the rank of captain, serving on the Polish–German border.

He was determined to assimilate into Polish society. He had reverted to his original name, Lajbcygier, at the end of the War, but was 'asked by army personnel to become Mayevski again so that the Germans should not see my Jewish name. They preferred that the Germans did not know that I was Jewish in case they felt that the Poles were trying to humiliate them.' He also felt that this would shield his children from the growing anti-Semitism, and was nervous that his involvement in the AK would lead to questions about his loyalty.

Florian retired from the army in 1964 and was actively involved in local life in Slubice, serving as a local councillor and president of the Combatants' Association. All this came to an end with the 'anti-Zionist' campaign that was unleashed after the victory of Israel in the Six-Day War in June 1967. When he refused to denounce 'Israeli aggression', his situation became increasingly awkward. In 1969, he got permission to make a trip to Brussels to visit a Polish friend and did not return to Poland. He settled first in Israel, where he found a surviving relative, and obtained a job managing his cousin's hotel in Netanya. In 1973, he met and fell in love with a guest in the hotel, an Anglo-Jewish widow named Sylvia. They married and he moved to England where he lives to this day.

This is a moving book, by a man with a remarkable instinct for survival. In spite of the many tragedies and disruptions that have marked his life, he is remarkably optimistic and free from bitterness.

Antony Polonsky
Albert Abramson Professor of Holocaust Studies,
Brandeis University
and the United States Holocaust Memorial Museum

Facing the War

The drone was distant at first, rumbling behind the clatter of tins and the slap of dough as we baked tomorrow's bread. As the unmistakable noise of the bombers got louder we began to fear tomorrow might never come.

The wireless had told us the Germans had crossed the border, but that was several hundred kilometres away and we had not expected them to get so close to Skierniewice on this, the first, day of the Second World War: 1 September 1939.

Uncle Shulim and I went into the backyard and watched the sky, realising what we had long feared had now begun. The blackout made it eerily dark and sirens warned of the impending raid. Those who could, tried to ignore their fate, and restaurants and bars were full of Polish soldiers doing their best to enjoy themselves.

Unexpectedly, the streetlights came back on. Soldiers and officers ran into the street and started shooting at the lamps to black them out. Within minutes, we heard bombs exploding at the railway station, where the army had been spending the day loading equipment onto trains heading for the border. Skierniewice was an important army city, just an hour by train from Warsaw.

Even though we expected the rest of the city to be bombed, my uncle decided we should carry on baking as much bread as possible. He knew people would be frantic. The queue for bread started forming at about 4 a.m. A few hours later we opened and quickly sold out.

As we served, we talked to customers and neighbours and asked what they planned to do. Nearly everyone was getting ready to leave; they were frightened the bombers would return. We heard that a German who worked at the electricity power station was responsible for switching on the street-lights; he was caught by soldiers and shot.

Uncle Shulim, my father's brother, was married to my mother's sister, Basia. So two brothers had married two sisters. Shulim owned the bakery and employed a third sister's husband, Uncle Josel.

Uncle Shulim said we had better leave town too, the two uncles, their wives and six children, and myself. We packed some belongings and food – we had saved a few loaves for ourselves – locked up and walked out of the city, joining the exodus of thousands on the road. We headed east for Warsaw, away from the advancing Germans, avoiding main roads, which were full of army vehicles and fleeing cars.

That first night we slept on straw with strangers in barns. Farmers in the villages put out pictures of Jesus and Mary, resting them against their gates and fences, hoping they would protect them from the bombs.

On the second day, as we were walking along the road, we heard the sound of planes coming from the west. Everybody lay down flat in the fields. The planes came over low and started shooting at us. When the planes passed people arose from the ground and looked to see who was killed and wounded. The screaming was never ending.

It brought home to me for the first but not last time the cruelty and reality of a war where civilians were considered as much targets as soldiers. Within a week, the Germans arrived at Skierniewice. We were taking refuge in a village about 30 kilometres away. We met people in cars and on horses who had been to Skierniewice and reported it was full of Germans. They said there was no point in carrying on to Warsaw as it was being bombed and would soon be occupied.

My uncles decided we might as well return home. We walked back and found everything as we had left it. We had a big store of flour, enough for a month. We opened up the bakery and carried on.

I helped out my uncles in the bakery but I earned my living as a carpenter. The following day I went back to Mr Broner, my boss in the carpentry workshop. He was finishing off some pieces. I asked him whether he wanted me to carry on working for him. He answered gloomily, 'No one will be needing furniture now. Go home, take care of yourself, I'll manage the best I can.' I went back to my uncles and became a full-time baker.

Skierniewice was full of Germans stopping people in the street and taking them to a square within a large park. There everyone was searched and their documents checked. On one man they found a pistol. They took him aside and shot him. Signs warned that all citizens had to surrender their wirelesses at a given place. I buried our wireless in the shed.

My cousins and I went to visit friends and exchange news. Some friends had not returned after running off and we discussed what might have happened to them. We heard that most of the country was occupied by the Germans. It wasn't until 17 September that the Soviet troops marched in and occupied about a third of Poland east of the River Bug.

We were visited in the bakery by officers who brought with them geese for baking in our ovens. Some of the soldiers were quite friendly. One officer, realising we were apprehensive, said to Uncle Shulim that not all Germans were Nazis, and that he was not happy about being in Poland, but he had been forced to come. He said he knew that Germany would lose the War. He knew we were Jews and hoped he could trust my uncle to keep what he was saying as confidential.

A couple of months after the outbreak of war, I received a letter from my mother in Sulejow informing me that my family were alive and well, but the house and shop had been bombed and burned down. German planes had flown over Sulejow when it was full of Polish soldiers. A soldier had fired at the planes with his rifle. The planes turned back and attacked the town. When they heard the bombing my mother, sister and brother ran out of the house and into the forest just outside of town. My brother had been ill in bed and had to flee in his pyjamas. My other relatives in Sulejow were unharmed. Most of the town ended up in flames. My mother asked me to stay with my uncles, where she believed I would be safer.

Childhood Lessons

I had left Sulejow in June 1939. I was born there on 4 May 1922 as Moshe Aaron Lajbcygier. Sulejow was a small town in central Poland in the province of Lodz with a population of around 6,000, about half of whom were Jews. They lived mostly on the east bank of the River Pilica where the town's sole synagogue was situated. The only church was on the predominantly Catholic west bank. Sulejow had two limestone factories and two wood mills, all owned by Jews, the main limestone quarry was owned by a Pole. Jews dominated traditional crafts like tailors, shoemakers and barbers.

My father, Josef, had his own bakery. He died of pleurisy at the age of 36 in 1924, when I was two years old. My mother, Rachel, had to sell the business and move in with her parents. My grandfather, Jakub Swierc, was a shoemaker and helped support us. We were not Orthodox Jews though we kept a Jewish home. Grandfather made kiddush every Friday night and we went with him to the synagogue every Saturday.

At Yom Kippur, the imposing entrance hall of the synagogue was full of large boxes of sand placed on tables. Members of the congregation would bring candles and light them on Erev Yom Kippur in memory of the dead. The heat and light of the candles generated a warm emotional glow.

I was always excited at the arrival of the Succos festival. Every year, from the age of ten, I built a succah behind our house for family and relatives. I used my grandfather's tools, bought strips of wood from the mill and erected the canopy covering it with reeds. We ate our meals inside it.

Grandfather was an admirer of Trotsky. He kept a large photograph of him on his bedroom wall. To him Trotsky was not a communist, but a social democrat.

I belonged to Skif, the youth movement of Bund, the social-democratic Jewish Socialist Party. I went to their hall for gymnastics and to act in Yiddish plays.

My mother started her own business buying fresh farm produce and selling it in Lodz. The Rosenbergs lived there – my mother's older sister Rajzel, her husband and their five children.

I started primary school, like most other Polish children, when I was seven. I went to register myself because my mother was away working. I finished my education when I was 14 in 1936. We could not afford to send me to college.

The town had two schools, one in the east, where mostly Jews went, and one in the west, mainly full of Catholics. I went to the school in the west where the Jews and Catholics were always fighting. Most of the time the Jews got beaten up, but I was one of the few who fought back and earned respect. One day I was attacked by three boys – I had suggested to the teacher that we fill in the space at the bottom of the fence that surrounded the school playground with earth, turf and flowers. The teacher said I could be responsible for organizing it. The boys were unhappy that I was responsible for giving them some hard work. They jumped on me, but I gave two of them a bloody nose. Blaming me for starting the fight, they went running to the schoolmaster, who pulled me out in front of the class and as a painful punishment forced me to kneel down on a bag of dried peas.

When I got home I told my grandmother, Bube Zisel, what had happened. She said she would scratch out the teacher's eyes. The next day she went with me to school. She was a tiny lady but she told off the teacher, saying Jews were not allowed to kneel, and asked why he hadn't listened to my side of the story. He apologized and said it wouldn't happen again.

One day a year, on Shrove Tuesday, the day before the start of Lent, it was traditional at our school to wipe all the blackboards with pig grease. Teachers were unable to use their chalk on the blackboards so all teaching was oral. It was known as Fat Tuesday. During the break, the Catholic boys would catch Jewish boys and force cooked pork sausages into their mouths, which would be spat out. They tried it on me. I opened my mouth, bit the sausage and, to their surprise, asked for more.

There were only three Jewish boys in my class and when they taught the Catholic religion we would go out into the playground. When the weather was bad, as I had a long way to go home, I would stay in the class and listen. I found it unbelievable that you could not even ask the question 'Why?'; it had to be how the priest said and there was no discussion. During one lesson the priest explained that Christ lived for 33 years, which was why priests had 33 buttons on their cassocks.

Once I had a fight with boys who tried to get me to stay for the religion class when I wanted to leave. They started pushing me, so I asked if they were such good Catholics did they know how many buttons a priest had on his cassock. The lesson in which we had learnt this had been a long time before and they had forgotten the answer, so I told them. That shut them up and they let me go.

My best friend was Mirek Kuczkowski, who was not Jewish. We started school on the same day. His parents had a farm where my uncle rented a flat. At weekends Mirek and I and his sister, Lotka, would take their cows to the common. Their mother would pack us boiled eggs and bread and butter.

Mirek and I liked to play in his father's barns. One day as we walked home I felt something in the lining of my coat going up and down my back. I took off my coat and stepped on it. When I shook it a dead mouse fell out.

Ever since the age of about 12 I had taken on the role of head of the family. My grandparents were elderly and needed to be looked after, and my mother was often not at home, so I did the shopping and cooking.

In the autumn, I worked out that it would be cheaper to store some vegetables through winter; it would also save me having to do so much shopping. I negotiated prices with the market farmers and got them to deliver to our house. My grandparents taught me the best way to store vegetables. Carrots and beetroots were put in sand in the cellar, and potatoes were kept on their own in the corner. Onions had to be stored in the loft where they could keep dry.

During the summer holidays, I helped my mother deliver produce to holidaymakers at their villas in the forest five kilometres outside Sulejow. They were mostly Jews from Lodz.

When I finished school I was keen to help take care of my mother, my sister Sara, who was working as a clothes shop assistant in Lodz, and my younger brother, Szulim, who was still at school. I decided to take up a profession and began a two-year apprenticeship as a carpenter. It was during that time that my grandparents died. My brother, my grandfather and I slept in the same bedroom. Some nights I liked to leave my bed and tuck under my grandfather's bed covers. One morning I told him to wake up but he didn't move. When I touched him he was cold, he had died in the night. My mother was in Lodz on business so I went to my Uncle Mendel who came and sorted things out.

There was growing anti-Semitism in Poland instigated by Endecja, the National Party. They called for a boycott of Jewish businesses and produce. They destroyed Jewish stalls in the market in Sulejow and beat up Jews. Members of left-wing parties joined together to help the Jews.

When I qualified as a carpenter in 1938 my boss, Mr Rosencweig, told me I had to look for another job because he could not afford to pay me a full wage. The economy was in a bad way and I found it hard to get work. I went looking in Lodz but had no luck. Eventually I found a job as a carpenter in Piotrkow, about 15 kilometres from Sulejow. I had some friends there I could stay with and was fed by my employer.

He would eat with his wife in the kitchen. She brought me my food in the workshop where I sat alone on a bench and ate. Over a lunch of borscht, mashed potatoes and a mince-meat cutlet, I started to cry. The food was tasteless and I was missing home. My tears dropped into my borscht, which tasted like salt water, not like my mother's.

On the first Friday I asked my boss for my wages. He said he would pay me the following week. I said I needed the money to give to my mother. He relented and paid up. I went home, gave my mother the money and stayed over Saturday.

Sunday morning my mother packed my sandwiches and I went to the bus stop for Piotrkow. As I was waiting I realised how unhappy I was and decided not to continue my journey. Back home I explained the situation to my mother. She said I did not have to work in Piotrkow if I did not want to, I could help her instead.

There was a lot of talk about a coming war. We knew what was happening to the Jews in Germany. After the annexation of Austria and Czechoslovakia we knew Poland, with its large ethnic German population, would be next. During 1938 and 1939 there was a rise of nationalism in Poland, and with it came increased anti-Semitism. One of the most popular slogans of the time implored Poles to buy from their own, implying they should boycott Jewish businesses. In spite of this the whole population was afraid of an invasion.

Uncle Mendel, my mother's brother, had four sons, three of whom were married and had done their national service in the Polish Army. Mendel and his sons were left wing and belonged to the socialist Bund Party. They were extremely disappointed to hear of the non-aggression pact between von Ribbentrop and Molotov. They expected to soon be called up. The atmosphere was tense; people spent all their time listening to wirelesses. And people who did not have wirelesses sat outside houses that had them where they were put on the window ledges for everyone to hear.

I suggested that my mother write to Uncle Shulim in Skierniewice. She did, and asked if he could arrange a job for me as a carpenter or perhaps train me as a baker. He replied, saying he had arranged a carpentry job and I should come to Skierniewice and stay with them.

In June 1939, I arrived in Skierniewice. Uncle Shulim introduced me to my new employer, Mr Broner, who made high-class furniture. I had to become an apprentice on a small wage again, as he was going to teach me about veneers and French polishing.

I was very happy with the carpentry job. I did not need money for myself as my uncle and aunt looked after me. In addition to the small wage I earned as a carpenter I got tips when we delivered furniture. Once I got a tip worth a whole week's wages, a silver 10 zloty coin. I sent all my money to my mother. In my free time I helped my uncles and learned how to be a baker.

Protecting the Family

In March 1940, I decided to return home. I felt that as head of the family it was my duty to take care of everyone. I took the train from Skierniewice to Piotrkow; trains were still running fairly normally. I then took the small train on to Sulejow, which was now the last stop heading east before the River Pilica where the bridge had been bombed.

When I arrived I went to the address at which my family were staying. It was an emotional greeting. I saw how they were living in discomfort and asked my mother what exactly she had been doing for a living after the shop was bombed. She told me she was buying haberdashery in the town and selling it in the villages. With the money she earned she could buy food and had enough for her needs. I suggested that I would go with her the following day and help her with her business.

We walked from village to village and in the evening arrived in the village of Siucice, about 25 kilometres from Sulejow. It was risky walking about. Jews were not allowed to go anywhere. If we had been caught we would have been sent to a camp.

I told my mother that we would be better off renting a room from a farmer in Siucice and opening a small grocery 'shop'. It seemed safer there than Sulejow; it was away from the main routes and was less likely to be visited by German troops or the Polish police who were based in the next village of Skorkowice. We could get our supplies from Sulejow once a week. She thought it was a good idea.

I found a room with a farmer named Jagudzki at the end of the village who also worked as a night guard. I introduced myself as Moniek because Moshe was too difficult for Poles to pronounce. They called me Maniek, which was even easier to say.

Jagudzki had to report to the village mayor that he had Jewish lodgers. The mayor in turn reported this fact to the police. The village already had several Jewish families who had lived there from before the War, they included a butcher and a businessman who bought and sold farm produce. The only other newcomer was a shoemaker from Lodz.

Farmer Jagudzki was a poor smallholder with a wife and two young children. Their cottage had three rooms, a small one for the children, a dining room and kitchen where the parents also slept, and a large spare room used for storage, which they rented to us. They emptied the room and I cleaned it up, painted it, and built two double beds, one for my mother and sister and the other for my brother and me. I put in shelves to display our products.

The only other shop was at the opposite end of the village. They sold groceries and vodka but were not in competition as most of our products were different, and they sent customers to us for things they did not stock. We had haberdashery, needles and cotton, cigarettes, saccharin, and many other household items. Word soon got round the village about our shop, which did good business and I became very friendly with the villagers, and boys and girls of my age.

In the room where we lived was an ancient disused stove built from bricks and clay, which I took apart and rebuilt. When the farmers' wives came shopping and saw the new stove they asked who built it. I got several orders to build new stoves in village farms and to do odd repairs like straightening doors.

Farmer Jagudzki had only one cow, and some pigs, chickens and geese. The geese ran out of his broken fence into the neighbours' fields and caused damage by eating their seeds. I suggested to Jagudzki that he should borrow a pair of horses and a cart and I would go with him to the forest to collect wood so that I could build a new fence. Once it was erected I got orders from other farmers to build them new fences.

I helped Jagudzki thresh his corn, rye and wheat. When the other farmers saw how well I helped him they asked me to do the same for them. I was happy to oblige.

Irena, one of the girls I became friendly with, was looking after her father's cows on the common near Jagudzki's farm.

We sat down and talked and at mid-day she had to drive her cattle back to the farm for milking. She asked me to help and when we arrived at her farm she told me to wait in the kitchen while she milked the cows, she would bring me a fresh glass of milk. While she was busy I talked to her father, Mr Mamrot, who had a carpentry workbench in the kitchen. I saw that he was having trouble with his stomach because he stopped to take some medicine. I asked him to let me try to plane the wood. He laughed, saying, 'Your hands are too delicate, Maniek, they weren't built for carpentry.' I said I would still like to try and he relented. I took the plane, got in position by the bench, and began to work the wood. Mr Mamrot said, 'Maniek, you're a carpenter.' He suggested I work for him; he had a lot of jobs in other villages but was unable to fulfil them all because of his ill health. I agreed and began working for him also.

Business was going well and the farmers' wives were asking for extra goods, including honey. As no one in the area produced any I decided to visit my old friend, Marian Kowalski, who had delivered honey to my mother before the War.

I had met Marian in the summer of 1938 when I was working as a carpenter in Sulejow. After work, going home to Opoczynska Street, I saw him pushing a bicycle with two baskets full of bottles. He stopped and asked if I knew anyone who wanted to buy some honey. I thought maybe my mother would; she could sell it in Lodz. I invited him home. My mother tried a spoon of honey and bought it all. She paid the price Marian asked. He was very happy and asked if he could deliver any more. My mother said she would have more in a week. He had a new business contact. He delivered his products and we became good friends. He invited me to stay weekends at his home in Blogie. I became friendly with Willy and Rudi Kurtz, his German neighbours. Marian introduced me as Maniek from Sulejow whose mother was buying his honey. He did not mention that I was a Jew. He didn't talk much about Jews to me or make any disparaging comments other than to complain about customers, Jews and non-Jews, who always wanted to have things cheaper.

I borrowed a bicycle from one of my friends in Siucice and

went to Blogie, about 25 kilometres away. Marian and his family were surprised to see me again and made me very welcome. I arranged to do more business with Marian as he was still selling honey. Now he was also selling home-made vodka, which I also bought. He delivered his produce to us in Siucice.

Marian was in his early forties. His blond hair was going grey at the temples. He'd never married and had studied with the intention of becoming a priest, until he became disillusioned with Catholicism. The age gap was irrelevant to us. He was easy to talk to and we often had discussions about politics. He was happy that I shared his views and believed in the sort of democracy we had before the War and was opposed to communism.

As Marian and I became acquainted, he confided in me that he was in touch with the underground. He gave me news of what was happening around the world.

Occasionally, when I went to Sulejow for supplies, I would get a lift from farmers who went there for the weekly market. It was risky for them to help me because they would have been punished for transporting a Jew. They said they would say they had just picked me up and didn't know who I was. Most of the farmers treated me like one of them. They all said, 'You are different.'

Some farmers gave me produce in exchange for my labour. On my visits to Sulejow I would take with me eggs, butter and cheese, and sometimes meat from the kosher butcher for my uncle and aunt and four cousins.

As I was now busy, my brother and sister took over my job of getting supplies from Sulejow during the winter of 1940/41. One day, in February 1941, we received a letter from my aunt and uncle in Skierniewice that they had received a parcel from our cousins, the Rosenberg family in Lodz, containing their daughter's dowry. As there was now a Jewish ghetto in Skierniewice they thought the dowry would be safer with us.

I arranged with Marian to go to Skierniewice and we went by bus, changing a few times to complete the journey of about 200 kilometres. We arrived and found our way into the ghetto through a gap in a fence. I was shocked at the changes to Skierniewice since I had left less than a year before. This was the first time I had seen a ghetto and I was surprised to see the overcrowding and deprivations with which people lived. A wooden bridge linked the sides of a ghetto spanning a main road that Jews were not allowed to go on. I watched as Gentiles walked and rode down the street freely.

My aunt and uncle had been forced to move from their bakery to a bakery in the ghetto. They had to live with my other uncle and his family in the same flat above the bakery. The family were apprehensive about what would happen to them; they realised the ghetto would probably be liquidated but could see no way out. It was impossible for a large family to escape. Marian and I stayed overnight and early the next day we took a sack containing the dowry of bed linen, tablecloths and other items and left the ghetto the same way we got in. The fences were not heavily guarded.

We arrived by bus in Tomaszow Mazowiecki, from where we had to then walk the remaining 15 kilometres to Blogie. We stopped for a snack at a bar on the edge of town. While we were eating, two German soldiers came in with rifles slung over their shoulders. They looked around and came to our table, where one of them kicked my sack. Finding it to be soft he lost interest. They turned and walked out. It brought home to me how dangerous the situation was.

We approached Blogie late at night, walking through the snow-covered forest. We came to a halt in a valley where we found the snow was covering a layer of melted snow. There was no other way of getting through, so we had to take off our shoes and socks and roll up our trousers and wade in up to our knees. Once through, we put our shoes back on and ran as fast as we could back to Blogie.

I stayed overnight with Marian and the next day went back to my family in Siucice with my sack on my back. A couple of months later we got a letter from my uncle saying they had made their way to Przyglow, a holiday resort five kilometres from Sulejow where I had gone to sell groceries with my mother. The family wanted me to return the dowry sack because they needed to sell it to get food. They were now just 25 kilometres from Siucice and I walked all the way with the sack. They had no bakery now and lived on their savings and by selling belongings. They were in low spirits. I suggested they joined us in Siucice, but they refused. They hoped to survive somehow, feeling that because Przyglow was in a forest off the main track the Germans might avoid it. It was the last time I saw or heard of them.

That spring of 1941 my mother, sister, brother and I moved out of the Jagudzki home; they needed the space for their growing children. We moved to a bigger room at the Morawska farm and ran our shop from there.

In April, Marian, who was still delivering his vodka and honey, brought with him a typed sheet of paper with news from the BBC that the Germans were preparing for war against Russia. It was a surprise because of the Soviet–German pact. We were hopeful that this would mean the Russians would drive out the Germans from Poland, forcing the Germans to leave the Jews in peace.

Death and Arrest

One day, I found out that my mother had been fined by the police; and, as we had no money, she was sentenced to a week in jail. I can't remember what it was for, but it may have been for the illegal sale of home-made vodka. I arranged that I should go in her place. I reported to the police station near Zarnow, about ten kilometres from Siucice, and I was put in a cell with about five Poles. We spent a lot of time discussing the War. They knew my name was Maniek, though nothing was said about me being a Jew. As I was the youngest, they were all quite kind to me and we became friends. I didn't find it hard to adjust to this new situation.

My brother, Szulim, was a quiet, studious boy. He was always reading newspapers and history books. He was very sensitive and serious and preferred to listen rather than speak. He was not as keen as I was to mix with other boys and girls and preferred his own company. But if my mother asked him to do something he was always willing to help out.

Szulim was due to go to Sulejow the next day to deliver some veal and groceries for relatives and to do some shopping. He took with him the note to show my left-wing cousins, who would be pleased to hear the news. Just before Sulejow, he was stopped by German soldiers and Polish police: Jews were not allowed to leave their towns and villages, nor were they allowed to slaughter animals or trade in meat, which was all reserved for Germans. The authorities set up ambushes, aware that Jews were flouting the regulations. When my brother was caught and arrested he was taken to Tomaszow where he was interrogated by the Gestapo. They found the note and wanted to know who had a wireless. During the interrogation Szulim bit an officer's hand and was shot. He was just 17. Marian found out the news from a police contact. He broke the news to me when he brought his next delivery and I had to tell my mother. It was all the more shocking being the first death in our family due to the War.

My mother was broken-hearted. She forbade us from going to Sulejow any more. We sold off all our goods in the store and shut down the business. Marian heard through the BBC that Jews were being transported to concentration camps. We were convinced the Nazis were going to wipe out the Jews, but we could not see any way out and waited for a miracle.

In the middle of June, I was arrested by a policeman in the village; I was walking in the street when he came up to me with his rifle and told me I had to go with him. No explanation was given. When I asked 'Why?', he told me he had to do it. I thought it must be related to my brother's arrest.

As he marched me through the streets my mother came running up to us. Someone had seen me being arrested and told her. She was crying. The policeman allowed her to say goodbye. She kissed me as I was led off.

I was taken to the police station in Skorkowice, where I was locked in a room with another Jew. I was sure we were going to be taken to a concentration camp and that my mother would lose her remaining son. The police were friendly; they apologized, but said they had been ordered to deliver a contingent of young Jews. The next day, we were put on a bench on the back of a cart. A policeman guarded us from behind. We were taken to Opoczno where we were handed over to the Germans.

We were assembled with hundreds of other Jews in a square lined with German soldiers. They marched us to the railway station where a passenger train was waiting. We were told to get on. I thought maybe I should try and jump off the train, but looking out the window I saw soldiers all along the platform.

We left Opoczno station around mid-day and arrived in Deblin as evening was falling. We were marched five kilometres from the station to the air force base, where a labour camp was still under construction. We were placed in a barracks with long slats of wood for beds. We had not been fed or watered all day and had to bed down without any blankets. We had been told we would be put to work the next day.

At 5 a.m. we were woken, given a portion of bread and a cup of hot coloured water they said was tea. At 6 a.m. we were marched three abreast to the airfield that was being extended. We were given shovels, spades and barrows and dug up the earth and moved it so that the ground was even.

At lunchtime, a lorry delivered soup and a piece of bread. While we worked we were surrounded by armed soldiers and given instructions by engineers. Everything was calm and ordered. We were not allowed to sit and talk and were treated civilly. Polish paid-labourers were building more barracks and we could see them go in and out freely.

We carried on working until 6 p.m. After returning to the barracks we got another piece of bread and some tea. We were able to wash and move around outside the barracks and chat to each other.

I observed what I could of the camp layout. We were not far from the entrance. More barracks were being built in rows stretching away from the entrance. After about a week, I decided I would try and join a group of four or five Polish labourers as they made their way out of the camp. I noticed the sentry on the gate did not check their papers and just relied on recognising them.

The night before I planned to escape I could hear the sound of heavy guns. It was 21 June 1941, the day the Germans attacked the Soviet Union. The border was about 100 kilometres away.

Early in the morning, before we were due to get ready to march to work, I watched for the Polish labourers who passed our barracks on their way to the gate after their night shift. I came out and walked alongside them as if I was one of them. They made no comment. As we passed the sentry I looked straight ahead. Inside I was trembling but showed nothing. I was dressed just like the labourers, so I did not look out of place. The labourers must have realised what I was doing, but they did not say a word.

Once outside the gates, the labourers went towards the village and I went straight ahead in the direction of the River Wieprz. The airfield and camp were between two rivers, the

Wieprz and the Wisla. I walked for a kilometre and followed a track to the river. On reaching the riverbank I saw a ferryman in his rowing boat and asked him to take me across. I did not have any money and he did not ask for any. I thanked him, saying, 'God will pay you,' an expression used by Christians.

From there I walked towards the road. A sign said 'Radom 70 kilometres', which was en route to Siucice, about 80 kilometres further on. My main thought was to get back to see my mother and sister and give them some relief from their suffering. I walked through the forest but kept close to the road. If I heard a car or lorry, I hid. There were large signs in Polish and German warning that entering the forest was punishable by death. The forest was used by the first partisans, groups of Polish soldiers who had refused to surrender at the start of the War. I pushed from my mind the thought that there might be mines.

I walked at a fast pace. I was afraid soldiers at the camp would soon find out I was missing and come after me. Once I had got far enough away from Deblin and felt safer I changed my pace of walking. In the forest I found a branch that I could use as a stick to support and defend myself. I slowed down and tried to look like a farm worker. Every few hours I took a break and sat down. I had no food but found streams where I could drink and rinse my face.

Most of the route was through the forest. I slowed down when I passed through villages. People passed me by but took no notice other than to say 'Hello'. Some women gave me the greeting, 'May Christ look after you,' and I responded likewise. To others I gave the same greeting to make them think I was one of them.

After the long forest came to an end and the fields began I risked walking along the road. I came across farmers, but no vehicles or soldiers. I arrived in Radom at about 10 o'clock having covered the 70 kilometres in about 16 hours.

In Radom, I saw signs for the Jewish ghetto warning that it was forbidden to enter or leave. I looked for a way to get in and found some loose wooden panels in a fence and sneaked through. On the street, I met people talking Yiddish and asked a couple if they could put me up. I told them I had run away from a labour camp and was going home. They fed me and let me stay the night. I was exhausted and fell into a deep sleep.

Early in the morning, I slipped out of the ghetto the same way I had come and joined the road to Opoczno. There was no more forest, just fields and villages. I was so thirsty in one village I greeted a woman in the Christian way and asked if I could have some water. She took me home and gave me water and a piece of bread. I explained that I was looking for work. I thanked her saying, 'God will pay you,' and went on my way.

That day I only managed 40 kilometres, and found a barn where I slept on straw. I had to cover 40 kilometres more and the next day skirted round Opoczno and arrived at Siucice late in the evening. I walked up to our front door and knocked. My mother opened the door and fell on me, almost fainting. She could not stop crying and said she did not believe she would see me alive again. She was sure she had lost both her sons. My sister, my mother and myself all ended up crying.

My sister said they had been told that in a couple of weeks all Jews would have to move out of Siucice to Skorkowice, where there were more Jewish families. The Polish police had orders from the Germans to concentrate all the Jews from surrounding villages in Skorkowice. We were well aware that this was a plan by the Germans to make it easier to eventually take all the Jews away. We thought maybe to ghettos, we did not suspect that it would be to concentration camps. We were sure the Russians would soon defeat the Germans and that the Jews would be left in peace. We had no choice but to take things day by day.

I had to make sure I was not spotted by the police. I told my young Polish friends about my escape, I knew I could trust them. They said they would warn me if they heard anything. My mother and sister were selling off their remaining stock of groceries. We lived off the proceeds so that I did not have to take on any work for the farmers other than helping Mrs Morawska with a few chores.

I found out that one of my friends, Jozef, was ill with typhoid. The entire village population avoided walking past the house, and only his mother and the doctor tended to him. I went to visit him. His mother warned me not to get too close, but I went to his bedside anyway. I gave him a drink and wiped his sweating face. His mother crossed herself, frightened that I might be infected. I was not bothered, though I was careful about hygiene.

Jozef knew I had been taken away and was pleased to see I had escaped. He knew he was going to die. He said there was no hope for him and died a few days after my first visit.I attended the funeral. The village knew I had been to see him and what close friends we had been, even though I was a Jew and he a Catholic. The day after his funeral, I went to the woods and cut down a young birch tree and made him a cross. I chiselled a hole and put in his picture framed by glass. People from the village went to the cemetery to view the cross, made by a Jew.

The appointed day came for our move. We packed our things and said goodbye to Mrs Morawska and other villagers and made our own way to Skorkowice, where we arranged to stay with a Jewish family; a butcher, his wife and two daughters. There was plenty of room in their large house. We were not allowed to leave Skorkowice, so I sneaked through the fields to visit Siucice, which was only a few kilometres away. I went to see friends and bought food.

I knew a tailor from Lodz who was now in Skorkowice and went to visit him. While I was there, a policeman came into his house for some repairs. It was the same policeman that had arrested me and taken me to Opoczno. He recognised me and said, 'You're back?' I said, 'Yes, I escaped.' I wasn't afraid because he had no weapon and if he'd tried to arrest me I would have run off. But he said, 'I knew you'd run away.' He never came after me, but I became extra cautious.

After a couple of weeks in Skorkowice, I got up early one morning to go to a farm outside the village. I got there at 5 a.m. to make sure I was first in the queue to use the grinding stones. I had some rye I bought from a farmer so that I could bake bread. I worked until about 9 a.m. and went on my way back with my flour. On the street, neighbours warned me not to go home; the police and German soldiers had taken all the Jews to Zarnow, about five kilometres away.

Losing my Family

I was shocked that I had been left all alone with no mother, sister or other Jews. I was torn as to what I should do. I didn't know whether to go into hiding or try and find my mother and sister. I gave the flour to my neighbour: I wasn't going to have time to bake bread. I made my way back to the house. No one had been given any time to sort out their belongings; they had to pack what they could in a few minutes.

The house was uncomfortably deserted. Everything was as they had left it, the beds unmade, the kitchen things unwashed from the night before. I resolved to go to Zarnow despite the danger, to be with my family. I stayed in the house all day. I gave the bedding to the neighbour and told her if we needed it again we would come to collect it. She said I could have it any time.

At sundown I made my way to Zarnow. The streets were deserted. Looking from house to house I found my mother and sister staying with yet another Jewish family. It was another emotional reunion. My mother said I should have stayed away. I told her that I needed to know what had become of them; I felt it my duty to ensure their survival. We had heard that the previous week, all the Jews from Sulejow had been transported to Piotrkow. We were aware that before long they would be coming for us.

I suggested we go back to Siucice and try to find a place to hide. My mother said we should not put other peoples lives at risk for helping us. 'We have no choice,' she said. 'Whatever happens to everybody else is going to happen to us.'

I told her maybe we could find a way to avoid such a fate. I needed time to work something out. I knew I couldn't just walk into the forest with my mother and sister and tried to think who might be willing to hide us. We finally managed to lay down well after midnight.

Early the next morning, German troops surrounded Zarnow and the German soldiers and Polish policemen went from home to home shouting, '*Juden raus*' [all Jews had to get out]. Soldiers with guns were everywhere, so nobody was able to even contemplate leaving the town. Poles watched us as we were herded into the square, where there was usually a market and cattle were sold for slaughter.

Soon the square was full of Jews of all ages. Frightened children clung to their mothers as if looking for shelter from these horrifying soldiers in helmets with guns in their hands. Old people followed their families as quickly as they could so as not to get lost in the crowd. We were all aware that we were being taken away to die. Some Jews did not want to leave their homes and they were killed instantly. The same happened to cripples and anyone who could not move quickly enough.

Everyone who was rushed into the market square was praying for a miracle to save them. There we all stood, crowded like a herd of sheep. One officer in an SS uniform shouted, '*Fachmene austreiten*' [tradesmen should step out]. It was close enough to Yiddish for us all to understand.

Everyone thought that it was another sadistic idea from the Germans. A kind of game they sometimes indulged in. He repeated his shout. He explained that workers would stay to work.

I whispered to my mother, 'Should I step out?'

'Yes my son,' she said. 'Step out.'

My sister agreed, 'Maybe one of our family will survive. I will look after mother.'

I said, 'Who knows if we will see each other again.'

'Think of yourself,' my mother said. 'We are already...' She was unable to finish her sentence.

I stepped out and a German tore me away from my mother. I just managed to wave my hand at my sister. She was looking at me with her big eyes without any expression in them.

The Germans took us 15 men and ordered us to pick up the dead, about 30 bodies, put them on a cart, and wheel them to the Jewish cemetery. The roads were stained with the blood of

Jews. We filled one cart mostly with elderly people and pushed it to the cemetery where we dug a mass grave, laying the bodies carefully down one by one next to each other. When we came back to fill the cart again, no one was left in the square. We took the second load to the cemetery, laid the bodies down and buried them all. It was late September 1941. The leaves were changing colour and falling on the ground.

Despair Followed by Hope

In a town of some 2,000 Jewish inhabitants only us 15 were left, including the leader of the Jewish council. He took us to a house and told us we could stay there.

'How about food?' someone asked.

'Find what you can,' he answered.

I did not feel like eating. It felt like my stomach's functions had stopped working that day while burying the bodies. I doubted I would ever feel hungry again. I imagined it would be good to dig a hiding place deep, deep in the forest, under the roots of trees, wrap myself up in a blanket, and sleep, sleep, sleep like a mole. I stretched out on the bed. I felt a rumbling in my head, ringing in my ears and a painful cramp in my throat.

I was tormented with thoughts. 'Where are you my mother, my sister? What will they do with you? Where have they taken you? Why am I here? How could I leave my mother and sister, two weak people? Why did I stay behind? Why? To save my own skin, my own life? Is it right to act like this when thousands perish? Can one think about oneself? What would I have achieved if I went with them? Perhaps I could have organized a revolt, an escape? Escape? Where to? Who would have taken us in and helped us if hiding a Jew was punishable by death? What should I do?

'Be quiet. I mustn't get excited. I must find a way out. I must not give up. Never give up. Maybe I should follow them? Try and find a farmer to stay with to be near them. Maybe take them some food. But where? But where? Where should I go?'

I remembered my question to my mother. 'Should I step out?' In my mind she answered me. 'Yes my child. Maybe you will manage.' And the kiss, the last kiss. The cold lips. As if I already felt death in her.

What did she mean by, 'Perhaps you will manage'? To stay alive; of course, yes, that was her deepest longing: but to stay alive, and not to lay silent in some hole. Somebody should and must remain. If only to remember! If not, who will ever be able to tell the story? Who will pass it on to future generations?

'My sister – forgive me for not taking you with me. Your eyes – the last look of a sheep under a butcher's knife – I shall never forget it.

'Isn't it a crime, what I did? Aren't I like a captain abandoning ship? Wasn't I the one who ran away first like a miserable rat?'

I was tearing out my hair, scratching my face. If I could bite myself, tear myself to pieces, perhaps that would ease my pain. I suffered all night and could not lie in one position. I was in an unbearable sweat and found the air stifling. I woke up every minute as each nightmare hit me.

Finally, I got up in the morning with a terrible headache, drained of any desire to move or live. Eating was out of the question. The leader came to our quarters and gave us axes and told us to follow him. I thought it was to do some road works, but he took us to a nice flat with modern furniture. We were to tear up the floorboards and look for buried valuables. He had orders from the Germans to rip up Jewish homes and gather up anything of value.

He took us to several houses. We searched the attics and cellars, tearing up floors, looking for hiding places. We took any objects we found and took them to his home. He locked them in a room to which only he had the key.

After the Germans got rid of the Jews from Zarnow they published a brief communiqué. 'Looters will be shot.' This was to discourage the Polish population from taking any Jewish possessions. The German soldiers loaded lorries with the valuables, including furniture. What they did not want, they distributed to the *Volksdeutsches* (the ethnic Germans); then the Germans left town, leaving the Polish police in charge.

I hated the sight of the Jewish leader, and doing his work, but there was no choice. I had to do what they ordered, and maybe the leader had to do it as well. We found no treasure under the floorboards, but a lot of hidden materials: leather for shoes at shoemakers, new clothes at tailors, shoes and other bits and pieces.

We spent a few days sorting it all out. After that we could do whatever we liked, as long as we didn't leave town. We had to find food for ourselves. The local men knew where to look for things, but I felt like an alien. So I bought some flour and started baking bread and selling it to cafes and bars. The flour cost me 100 zlotys and in a few days I managed to treble it.

As a carpenter, I was sent to the police station to work for the police chief who lived there. I was told to chop logs of wood into small pieces in preparation for winter and store it in the shed. I cut the logs to an exact measurement and chopped them, laying them in neat rows in the shed, filling it up to the roof.

As I was cutting wood I began to plan how to run away. I could not see any way out of this situation. I knew that any day, a lorry with Germans could appear and take us to some place from which nobody had returned. I could let out my anger only on the block of wood that I was hitting with all my strength. I did not notice that the young wife of the police chief was approaching me.

'Be careful,' she said. 'You'll hurt yourself.' I did not answer and did not look at her. 'You don't look at all Jewish.' She tried to make conversation.

'So what?' I replied. I did not stop working.

'You look more like a village boy.'

'That's what I am.'

'Not a Jew?'

'A Jew living in a village.'

'Really?'

'Yes.'

'Where?'

'In Siucice.'

'I come from Skorkowice. That's not far.'

She seemed lost in thought. I had no idea about what, but I felt grateful for the little sympathy I saw in her eyes. In those days, a little sympathy was more valuable than vodka;

sweeter than sugar. She walked away with her head down and told me to come up to the kitchen. Perhaps she did not want me to see her tears.

It was quite early to call me for lunch. When I entered the kitchen she poured hot water into a bowl and persuaded me to have a good wash. I felt embarrassed to take off my shirt. I hesitated and finally gave way. Refreshed, I felt better and sat at the table and ate with a healthy appetite for the first time in many days.

She served up bread and butter, cheese and jam, boiled eggs and coffee with hot milk. On the windowsill I noticed a packet wrapped in newspaper. I thought it must be some bread and butter for me like she had given before.

'Eat, eat, as much as you can,' she was telling me again and again. Just like my mother, I thought. Suddenly, I could not swallow. I could no longer eat, only drink. She was surprised.

'Can I take the eggs with me,' I asked.

She pointed to the parcel and said, 'And that as well. But eat some more. You've eaten so little.'

'I can't eat any more.' I asked if I should come to the house the next day.

'Yes, yes, come. You can help.' She gave me another piece of bread. 'This is for your friends,' she said. Just then I heard heavy footsteps. It was the chief of police.

'Look at this boy,' she said to him as he took off his cap. 'He comes from a village not far from where I was born. He doesn't look Jewish and speaks very good Polish.'

'I know. So what?'

'Can't you do something for him, help him?'

'Help? How? Who can help him? I've got orders to keep my staff at the station tomorrow. Something is going to happen. Probably they will take them away.'

'Where to?' she asked.

'He knows where to; where his mother and sister were taken.' The policeman looked me up and down as if to say, is he pretending to be an idiot, or does he really not know? 'Even if it is where your mother was taken, you will not see her, anyway.' He did not need to say anything more. I rushed out, forgetting to take the food parcel. The good woman was shouting something after me but I could not make it out. I ran then walked; walked as if drunk, deaf and blind to anything around me.

Mother, sister – they are all no longer alive. I fell on my bed crying. At first, I could not speak to my friends who surrounded me, bringing some water, not knowing how to help me.

'Our people are dead,' I managed finally.

'How do you know?' They all wanted to know.

'From the chief of police. He said the Germans are coming to Zarnow tomorrow, probably for us.'

Isaac the shoemaker jumped up and rushed out. He had some savings and was planning to hide at a farmer's house somewhere in Skorkowice, the village where he was born. The others disappeared from the room and everybody went off on their own to work out what they wanted to do. Some said, 'What will be, will be.' Others planned how best to run away.

I decided to leave late in the evening. Suddenly, Isaac came into the room saying there was no hope for us. He told me that he asked many farmers to hide him, but no one wanted to take the risk. I asked him how he had come back from Skorkowice.

He said, 'I've also been to Siucice, and Marcin brought me back on his horse and cart.'

I asked him where Marcin was. When he told me, I ran out to see Marcin, who I knew very well. He was pleased to see me and greeted me like an old friend. He told me that no one wanted to help Isaac, even with his money. No one wanted to take the risk. 'But you are different.'

I asked him if he could take me back to Siucice. He said, 'Yes. With you it's another matter. Anyone would be willing to hide you. The Germans couldn't see the difference between you and our boys. On top of that, you can do anything. You know all about work in the fields and the farms and carpentry. You can even stay for a while with me. I need to have some frames made for the windows and doors for my new house.'

I could not believe my ears. Was it possible? I, who had nothing, was worth more than Isaac with his money? There was not much time for reflection.

'You have nothing to take with you?' Marcin asked.

I understood what he was getting at. What he was doing for me was worth a lot. But what did I possess? Some underwear and the clothes I was wearing. I thought of the Jewish leader.

'Wait a minute,' I said to Marcin. 'I will be back soon.'

I ran to see the leader. I was thinking of the rooms full of leather and other valuables. Why should he not give me some of it? I was sure he would under such circumstances. I knocked at his door. He was about to go to bed, but he opened up (seeing that it was me and not some Germans). He shouted at me, 'Go to the devil. Don't disturb me when I'm about to go to sleep.'

'I'm not going anywhere until you hear me out.'

'What's this about then?'

'I can't speak about it on the doorstep.'

'Can't you come back tomorrow morning?'

'No. It must be right now.'

'Wait a minute.' He opened the door. He was wearing his dressing gown. I could see his hairy chest under his pyjamas.

'So what is it then? Quickly.'

'Tomorrow we are going to be deported.'

'How do you know?'

'From the police station. I want to run away.'

'Run,' he replied. 'Who is holding you back? Fly away like a bird.'

'I want to hide in the villages.'

'They will kill you.'

'Probably. But I will live for a bit.'

'Good luck,' he said.

'I need some money. I am poor.'

'Why are you bothering me? I'm poor too; that is why I am not running away. I am waiting for my turn. You should do the same. It will save you disappointment.'

'I want to live.'

'Who is stopping you?'

'Give me some leather, some of the material from the stock.'

'Crazy. You want me to be shot right away.'

'No one will know. A farmer is waiting for me. He will take it and help me.'

'Get out – and not a word about it. One piece of leather or material will give you nothing, and you'll still die anyway. And because of it, I'll be dead too.'

I slammed the door and went out into the darkness. In our quarters, I put my things into my rucksack and started saying goodbye to the friends I was leaving behind.

'Where are you going?' they said in chorus.

'Where my legs will take me.'

One part of me regretted not telling them where I was going. But I knew that if the Germans started asking questions it was better that no one should know.

Marcin was waiting for me patiently. He was disappointed to see my little bundle. However, he started the horses right away and we rode off.

Going into Hiding

I yawned and rubbed my burning eyes. I looked round about with anxiety. Where am I? I felt terrified. I was lying on straw covered with a blanket; there was a cushion filled with hay under my head. Above me hung a ceiling of straw and on every side stood the bare stonewalls of a stable. To one side lay some cows.

I tried to think clearly. 'Oh yes. Yesterday I was in Zarnow and during the night Marcin brought me to Siucice. Why has that journey not remained in my memory? Have I been so absorbed by recent events?'

I felt listless and empty. If a German had suddenly appeared and pointed a gun at my heart or head I would not have raised an eyelid. I closed my eyes. How good it would be to sleep months and years if the stomach did not demand food. I would gladly creep into a tomb and sleep, sleep, and sleep until the end of this nightmare. But one had to eat. And nobody would give me food without work.

Marcin had brought me here so that I would do a job for him. Carpentry. Making doors, windows, floors. He would have to pay quite a lot. Today he will provide me with just my food for the job. And he deserves all of it. God bless him. Who knows, perhaps he has saved my life. That's almost for sure.

In Zarnow I was condemned. Despite the Jewish leader refusing to help me, and despite having no money, I had managed to get away. Now I have a chance; one out of thousands. I have got this far – and then what? No point worrying now. Take it day by day; and another one, and another. And if it comes to the crunch, perhaps something else will come up for me out of the darkness. I have to grasp it like a man trying to keep afloat as long as possible on stormy seas. I have to see it through to the end, but what sort of end will it be?

I put my hand under my head and closed my eyes. The farmer's dog howled and came out of its kennel, clanking his chain and wagging his tail. I knew that someone was approaching. I got up carefully, and went to the tiny window to look out on the yard. The farmer was coming towards the stable. He calmed the dog. He was carrying a bucket, probably with some food for the cows.

A minute later, Marcin entered and took out of the bucket a jug with soup. From his jacket he took out a piece of brown bread and a spoon. He gave it to me. 'Eat,' he said.

'Thank you very much.'

'You eat and I'm going to look for some tools, a plane, a chisel and saw.'

'I could do with two planes, one to plane and one to finish off, and some different sized chisels.'

'I will try,' he said. 'And don't you come to the window. God forbid if someone should see you. You know what it would mean.'

'Don't worry.'

'I'm going. You eat; the soup will get cold. It's important you get some nourishment. What's happened has happened, it can't be helped.'

When he went I sat down and put the clay jug between my legs and ate my soup. I ate half of the bread and hid the rest, just in case.

I renewed my friendship with Burek the dog, who I had got to know when I had worked for Marcin before. I had come to the farm quite often, sitting and talking to Marcin's daughter, Helena, who had been taken to Germany as forced labour.

I stretched out on my bedding, hands under my head. I was looking towards the window. Sparrows and blackbirds were flying all around. Slowly, the village was waking up. Cocks were crowing, cows moaned, and dogs barked everywhere. The day's work was beginning. Only I – in the village where I had lived and knew everybody, where I had never harmed anybody – had to lay low and hide.

I had to suffer because of a faith that I no longer believed in. Like so many others, I felt that God had abandoned the Jews. I felt I had belonged in the village where I had many good friends, but now I had to isolate myself from them because a hoard of armed barbarians were ready to pounce on me. I was helpless. They had already murdered my brother and deported my mother and sister to their deaths. Only I remained.

I nodded off and slept for quite a while. Something kicked me in the foot. I sat up, frightened. No, it was not a black-uniformed Nazi or policeman. The farmer towered above me. 'Look what I brought.' He was displeased, saying, 'The moment you ask someone for anything they want to know for what, for whom, why. It seems they sniff you out in no time.' I knew quite well that villagers had a very keen nose. One look at the clouds will tell you tomorrow's weather. The slightest indent on moss will betray what animal passed that way. And, if a carpenter is borrowing tools, then he must surely have a worker in the house.

I asked Marcin if he had got the tools from Mamrot, the village carpenter.

'Yes,' he said. 'I am not afraid of him.'

He took out the tools from a sack and lay them down. I said I would start work straight away.

'May God protect you,' said Marcin as he crossed himself and scratched his head. He was probably already regretting his kind deed, but greed prevailed. I got myself ready and went over to the new house and began work.

I had to make some noise, which could easily attract the attention of the villagers. However, Marcin's house stood away from the main road and that helped a bit.

Some people thought that the farmer himself was working at the house. But others heard the noise when Marcin was in the village. Some curious people tried to peep in, but Burek kept them at bay. Days and weeks passed by. I worked non-stop from dawn to dusk. I tried not to think about the past or the future. After supper I fell asleep immediately. I slept deeply, content with this existence. I would gladly remain doing this sort of work until the end of the War. I became accustomed to the War, to the work and the solitude.

One evening, I had just fallen asleep when I felt someone pulling my arm. I woke up, startled. 'What is it?'

'Don't shout, it's me, Marcin,' he said, as he calmed me

down. 'I wanted to talk to you.'

'What about?'

'You know...' he started uncertainly. 'You finished a good chunk of work. I am very pleased. But I am very afraid. And my wife is even more so. Perhaps you should now go somewhere else, because people are resentful.'

I could see what he was trying to tell me. The slave had done his best over the three months he'd spent with him. Now only a little more work remained and he did not want to take any more risks. I felt very sad. Not that I was expecting any gratitude. But I felt sorry to leave a place so calm and quiet, not a living soul all day. And now, go. Where to? It was February, the coldest time of year. Everywhere was frozen and covered with snow.

'But nobody has seen me here?'

'But some heard you. Some were joking that I sold my soul to the devil for some carpentry work. Others said it must be haunted.'

'So what shall I do?'

'Tomorrow you will go to Walenty's barn. You can thrash some wheat for him. Then you can go to Stach, who also needs help.'

'Well it's better than working in a labour camp or getting sent to a concentration camp,' I said.

I did as Marcin suggested. Other farmers put me up in their stables, making a space for me to sleep, and I worked in their barns. In all, I worked on about 20 different farms. Sometimes, I would approach the farmers by going up to their houses at night, making sure only the family were there, and asking if I could stay. In some farms, I would go straight into the stable late at night and make myself a place to sleep. They would discover me in the morning, but never used to mind, and would give me food.

Late one night, I went in to Morawska's stable and found no straw to bed down on. I lay in a wooden trough used for feeding the cattle. I put in some hay but had nothing to cover me. I woke early and found my left arm and side was frozen stiff. I could not open my mouth or move the left side of my face. I started to rub myself and forced my mouth to open. I moved about and got my circulation running. I can still feel the consequences of the frostbite in my little left toe.

The sight of women and girls while I was moving around the village always made me think of my mother and sister. Where are they now? Are they still alive? Sometimes I felt like throwing up everything and running; running to look for them so that I could help them, or at least to put some flowers on their graves. But where were their graves?

There were terrible rumours. The farmers were saying that in Auschwitz they shot hundreds of people every day and burned their bodies. Nobody came out alive. And every day, new transports of Jews arrived. It was just a place of slaughter. And I was here, alive.

When it became quite clear to me that I would never see my mother and sister, I experienced a shocking sensation and started to cry. And, again, I felt like going there to see for myself. I could not believe that it was possible in the twentieth century in the centre of Europe. But I had no money. Nobody paid me a zloty for my work.

'Pay for what? You can eat, sleep; we do your washing, give you a shirt and trousers. Why do you need money?' said Piatkowski, one of the farmers.

It was true, what did I need to save money for? So it could be taken off me when the Gestapo came for me? As long as I was able to work they would feed me. I got used to this way of life and became less fearful of being discovered, so much so that I even went for walks in the evenings.

Winter was over and I was pleased to see the spring sunshine. I needed to get away from the smell of stables and farmyards and get some fresh air in the forest, on the hills of Diabla Gora, about ten kilometres from Siucice.

One evening, I left Siucice on my way to Skorkowice. Just before the village I went to the cemetery where Jozef was buried. I had attended many funerals there and knew which tombs were open. I went into one where three wooden coffins were standing on the concrete floor. It was quite cosy and warm. I lay down on a coffin and fell asleep. I woke up after a few hours and was unable to fall back to sleep because I was uncomfortable. I knew there was not enough time to pass through Skorkowice and reach the forest by morning, but I did not want to stay in the tomb all day and decided to go into

Skorkowice early in the morning, where I slipped into a barn and went up into the loft where there was plenty of straw.

I planned to stay there for the rest of the day and sleep. As I was making a space to sleep, I noticed that nearby the straw was moving. I moved towards it, moved the straw, and got a fright when I saw a man lying there. I recognised him as Jankel the baker, who was born in Skorkowice and had lived there all his life working with his father in the family bakery. We had baked bread together while staying in Zarnow. He was one of the 15 tradesmen who stepped forward.

I woke him up. He was startled but relaxed when he recognized me. I suggested that we should keep together and hide so that one could look out for the other. He did not think it was a good idea. He said he had some farmers willing to help him hide. His brother was also hiding separately in the village; he thought it was best that way. We stayed together that day, and in the evening I left and passed through Skorkowice on my way to Diabla Gora.

I stopped in a village just before the forest where I asked a farmer if he could give me any work. He said he needed work himself. All the local farmers were poor because the ground was not fertile. He gave me some food and I made my way to the forest.

Crossing the fields, I saw a mound of stored potatoes in the ground. I managed to uncover it with my hands and put some in my rucksack. The forest was hilly. I climbed up and found a gap between rocks that felt a safe place to stay during the day. I collected dry wood and made a fire. I put about 30 potatoes on the smouldering ashes. When they were baked, I cleaned off the ashes and ate them for lunch. Later in the afternoon I baked and ate another batch.

I had spotted some wooden planks stored in a field under a straw roof. Once it was dark, I went to check it out and realised it would be a good place to sleep, with plenty of straw for bedding. In the morning I returned to my spot in the forest. I carried on this routine for a week.

I collected some water in a bottle from the village which I rationed, using it as if it was medicine. On the rocks, I used a piece of stone to scratch 'Lajbcygier was here' and the date.

One evening when I was preparing to go to the village I

heard the noise of lorries. From a distance I could see German soldiers had stopped in the village. I went to my secret place in the fields under dark, and early in the morning before light returned to the forest. I was frightened to make a fire. After daybreak I went back to watch the village; it was still full of Germans who were shouting about something.

I made a fire and baked potatoes for lunch. Late in the afternoon the lorry engines started and the Germans moved away. Come the evening, I went back to my sleeping place. The following morning – making sure all was quiet – I went into the village on my way back to Siucice.

I asked a farmer if the Germans were still in the village. He said they had stayed for 24 hours before continuing on their way to the Russian front. They had taken plenty of food, chickens, butter and eggs from the farmers who were glad to see the back of them.

I made my way to Siucice, stopping at my friend Stefan's farm in Chorzew, away from the track on the outskirts of Siucice. His mother gave me a friendly greeting and invited me to stay for a few days. I slept most of the day in the stable and at night helped Stefan make vodka, which he sold to other farmers. It was illegal under the German occupation to make your own vodka and farms had been burned down because of it, though that did not stop many farmers taking a chance. The Germans wanted all grain production to be used for food.

After about a week I carried on to Siucice, where I stayed at different farms. At some farms I crept into their stables at night and revealed myself in the morning when someone came to see the animals. I was becoming increasingly unhappy about going into a stable without permission; I much preferred going first to the house and asking, but sometimes I had no choice if I thought that strangers were in the house. I got to know which of the farms around Siucice did not mind me staying briefly, so long as no one saw me come in.

The weather was getting warmer and I preferred to stay throughout the day in stable haylofts, where the air was fresher and I could remain warm. When I was invited into homes in the evening, farmers would often express the hope that the War might be over fairly soon. The Germans were advancing into Russia, but the farmers did not believe they would survive the next winter. They would share the same fate as Napoleon.

The farmers talked about partisans being active in parts of Poland. I wanted to join but did not know how to contact them. There was no question of any anti-German action in Siucice. Who would dare to confront such a heavily armed enemy? The village was far from the main road between Sulejow and Opoczno. The Germans avoided wooded areas and the uneven sandy tracks leading into Siucice. But the Germans were not taking any chances and guarded against any future partisan attacks by ordering farmers and the forest authorities to clear the woods about ten metres either side of the roads.

Life in Siucice was relatively free of the noise and panic of war. Old policemen and village elders functioned much as ever. They did their duty. No more. And for a bottle of vodka looked the other way. This superficially idyllic way of life kept up my spirits. It encouraged my belief that it might be possible to live through it all, and with it the hope that, one day, I would see the death of fascism and the dawn of a better order in Poland.

I spent the months between February 1942 (when I left Marcin's farm) until April (when I left Siucice) wandering between farms, spending a few days here or there, sometimes working for my keep and other times lying low in stables.

On some of my evening walks I met some of my old friends, who were pleased to see me, though they weren't surprised; they'd heard I was in the village and asked if I needed any help. No one asked me to come to their homes because they were frightened for their family's safety. Occasionally, they would warn me that a policeman was coming to the village.

My only upsetting incident in Siucice occurred when I visited a widow who had two sons in their thirties who were in business delivering flour to Sulejow. One evening, I turned up at the farm and found the woman alone. She was friendly and gave me bread, butter and milk. As I was eating, one of her sons came home and asked his mother what I was doing there. She said, 'The boy is hungry. I gave him something to eat.'

He came up to me and stood over me and said, 'Do you know I can take you to the police?'

I said, 'Yes, I know.'

His mother started to shout at him, 'Are you stupid? What do you want from him? Leave the boy alone. It's none of your business. He'll go when he's finished eating.'

So I finished eating and went off, thanking her. As it was late in the evening I went out of Siucice. In the middle of a field was a farm where friends lived. When I got close to the farm, I realised there was no light and they were all asleep. The dog was outside and barked when it heard me. I left and, not far away, found a high haystack. I was thinking whether to lie in the hay till morning and then go to the farm. They had put me up before. I pulled out some hay to make a cubby-hole and crawled in leg first. My head was at the edge and I covered it with a little bit of hay.

It was very cold and I could not fall asleep. I got out and started to exercise to warm myself up. Standing and watching the clear sky and the full moon I started to cry, saying, 'Mother, why did you leave me? What shall I do? You can't help me and I can't help you. But I can't give up. Remembering the words of my sister, "Maybe one branch will survive." And I will try to survive.'

After a good cry, I moved back into my place in the haystack and fell asleep. I woke early and heard noises; someone was taking water from the well. I moved into the farm where they let me stay during the day. In the evening, I moved back to Siucice.

Becoming a Hermit

As I was walking through Siucice late in the evening I came across Mr Jagudzki, who was on duty as night guard. He had been putting me up from time to time, just as he had helped out when my mother and sister stayed at his house and he let us sell haberdashery from our room. The Jagudzki house was the last in the village. I always felt safer there because from the kitchen window there was a clear view down the side street to the main road.

Mr Jagudzki looked concerned. 'Maniek,' he said, 'Be careful. I met the policeman, Sieczko, last night, who asked me if I knew about a Jewish boy hiding in the village. I told him I haven't heard anything. He wants me to let him know if I get any information.'

I was shocked. This was very bad news. After all I had been through in the winter, now I would have to move again. Jagudzki said, 'Go to villages where they don't know you. You could get work on a farm; you can turn yourself to anything. You're young and strong and you don't look Jewish. It would be a shame if you ended up in a crematorium.'

I said goodbye to my friend. A heaviness descended on me and my legs started to shake. In the next street I met some Polish friends who knew I was in hiding. They noticed my anxiety. We talked, but I do not remember what we said. I only had one thought in my mind: I had to go. But where, where?

At the end of the village I went into a farmer's stable where the dog knew me and licked me in greeting. I lay down on straw in the corner but could not sleep, turning from side to side. I dropped off briefly only to wake with a fright, convinced they were coming to get me. There was only one way out of the stable. I was ready for anything. If they came for me I would fight with a pitchfork. They're not going to take me alive. I'll die like my brother.

But no one had seen me go into the stable, and the dog stayed outside on guard. Finally I fell asleep, and woke again with a troubled heart. Too many people knew about me in the village.

I decided to wait in the stable till evening. I heard someone coming. It was Morawska, the farmer. A widow with four children, her two grown boys had been taken to Germany as forced labour. She still had her little boy and girl. I gave her a fright when I appeared out of the corner, but she knew me well; I had stayed in the stable many times. I told her only the dog had seen me come in and asked if I could stay till later.

'Yes,' she said. 'You can't go anywhere now. I'll milk the cows and bring you something to eat.' After finishing milking she went into the house, returning with a bucket containing a jug of soup and bread. I was famished, and as I ate, felt the life come back in to me.

The cows had to be let out to graze. 'You can go up to the loft,' said Morawska. 'I'll bring you something to eat at lunchtime.' Lying down on the hay, I contemplated my fate, and resolved there was only one choice: to go to Marian in Blogie. I knew the way well, but I had to be careful.

This is where I shall start a new chapter of my hiding life.

I looked out through the gaps in the loft. Life on the farm was going on as usual. Before dark, the cows came back from grazing and Morawska brought another bucket, with a jug of soup and bread for now and something for the journey. I thanked her and asked if I could pop in again some time. She said that was fine.

After dark I left. Most farmers were having their supper. I made my way behind the stable on to the road leading to the village of Ciechumin. In the day, there were many people moving between villages. At night all was silent. There were large woods between the villages and I preferred to walk through them. By now I was used to moving at night. The sudden noise of an owl gave me a fright. After some 15 kilometres I felt hungry and sat down to eat my bread. Hunger is a good sign, I thought. When the Germans took away my mother and sister I could not eat for days.

There had never been any Jews where I was heading. It

was mostly Polish farmers and a few farms owned by ethnic Germans. I continued my journey for a further ten kilometres or so, pondering whether Marian would help me and how his parents and sister would react.

I arrived in Blogie around midnight. The houses were all dark. I approached Marian's house and saw a lamp light between the curtains. I knocked on the window.

'Who is it?'

'It's me, Maniek.' Marian let me in. I explained that I had to leave Siucice because the policeman, Sieczko, was looking for me.

'I'll help you, but you can't stay here: the German family is still over the road and the police station is too close for comfort. The policemen still pop over for a drink.'

Mr Kurtz had been killed by a Polish villager on the first day of the War; Mrs Kurtz was left with their two sons and a daughter. Willy, my age, was forced to join the German Army. Rudi, two years younger, stayed at the farm.

'Maniek, you can stay overnight in the stable and in the morning, before first light, I'll take you to the forest and show you a place to stay. You'll have to stay in the forest until it's dark, then come through the back garden behind the church to the stable. Janina will bring you hot food. During the day, while you're in the forest, you can bake some potatoes.'

Janina heard us talking and came out of her bedroom. Marian explained my predicament. She agreed with our plan and gave me some bread and butter and barley coffee. Marian suggested we go and prepare a place to sleep in the stable. We took a torch and some old coats. It was still very cold. I made a bed of straw in a corner.

Marian left me in the dark. Tired as I was, I could not fall asleep. I was thinking about what would happen to me. Maybe Marian would give me away. I was sure he was trustworthy and I was willing to take the chance. It was getting colder, so I covered myself with the coats and finally fell asleep.

I had always got on well with Marian's sister, Janina, when I stayed with them on weekends before the War. I remembered always enjoying her boiled eggs, taken fresh from the hens. Marian wanted to see how many eggs I could manage to eat in one go. I took up the challenge and downed ten eggs with horseradish to help me digest them. They were amazed.

Janina was a very small and slim woman. Her blonde hair was always tied in a ponytail and she never went out without her scarf on her head. She was not very educated but reasonably intelligent. She had spent all her life at home taking care of her parents, cooking and cleaning and looking after the chickens and rabbits. She did seasonal work for farmers, weeding and gathering wheat and rye in the summer, and harvesting potatoes in the autumn.

I always thought she was not that keen on men because she never got on well with Marian. They always argued. He would shout at her, but she kept quiet and would try to calm him down.

She had a friendship with Franek; they would sit up and talk half the night, but it never led to romance. He'd suggested marriage but she didn't want to leave her parents. If she didn't have her family responsibilities, she said she might have become a nun.

I was the first Jew she had got to know. She had no opinion about Jews, and wasn't prejudiced. She treated me with the respect she gave everyone else.

Marian was frustrated with his life: he thought he deserved better. When he was young, he had thought about becoming a priest or a monk. But then he said he found out they didn't practice what they preached. He had studied for several years, but his love of Catholicism turned to hatred. He talked with anger about the Church and priests; of there being many homosexual priests. One or more had tried to seduce him. He said his neighbour, the priest, used his housekeepers and staff for sex, and had an illegitimate son who he provided for. Marian never talked about religion, though from what I understood, he believed that Christ was an invented figure. He would only go to church a couple of times a year, at Christmas or Easter, so he probably still had some faith. Janina went to church every day.

Marian never talked about women or sex, and if he had any love affairs, he never told me about them. He had a woman friend in the village that he sometimes stayed over with, but I only knew her to be a good friend.

In the village, Marian was treated with respect, as a man of intelligence. He got on well with the police. He had written a book about the history of a church in Podklasztorze, near Sulejow, and the surrounding region. It was published in the early 1930s when he was in his early thirties. He gave me a copy which I lost when our house was bombed. He was a frustrated writer who could never make a living out of it. His income came from his beloved beehives, where he spent most of his time, his smallholding and potato field, and selling rabbits for breeding.

It was still dark when something woke me, brushing my face. I opened my eyes to see a white rabbit sitting in front of me. I stroked its soft fur. It kicked its back legs and moved away, stopping to look at me. I noticed some light coming in through a small window and gaps in the door. Marian came in soon after.

'Get up,' he said. 'We have to go. It's getting late.'

We went out through the back garden towards the forest. I had some potatoes in my rucksack and matches in my pocket. I was frozen, but soon warmed up as we walked deep into the forest. About five kilometres in we reached the virgin forest.

On our way, Marian explained that I should look out for markers so that I could memorize the way in and out. By the road, I noticed some treeless gaps where there were once houses and which gap we had passed. I tried to remember which animal tracks we were taking, as there were no paths.

In the virgin forest the pine trees were in straight rows, very tall, thin and close together. Only small animals could have squeezed through, their tracks were everywhere. We had to bend down to avoid branches and managed to squeeze through to find a big enough gap so that we could sit down. Among the pines were clusters of white birch.

Marian stopped. 'You can stay here and make a fire, but only from dry pieces of wood that won't give off smoke.' I started collecting dry wood to make a fire. Marian helped out.

We sat down by the flames and Marian imparted his knowledge of the forest, how I could survive, how to get to the River Pilica five kilometres away to fetch water. He warned me to be wary of wild boar and never to go near a mother with piglets. If they attacked, the best thing to do was to jump up a tree.

'Now I have to go,' said Marian. 'Come back to the stable this evening. I'll get some tools for you.'

'Thank you, my friend,' I said. Marian left and I was alone.

I was frightened by the change and the loneliness. At Marian's home I had friends; here, I had nobody. Would I be secure here? I was sure of one thing, I would not be troubling anyone and risking their lives. Maybe Marian was thinking the same thing. At his home I was a threat to his entire family.

I stayed in the forest all day and after sunset made my way to Marian's stable. There I laid myself down in my spot. The rabbits came to me; I cuddled them and was happy to have their company. After a while, Janina came and brought a bucket containing a jug of soup and bread. The soup was comforting and the bread fresh.

'I have not had fresh bread for a long time,' I said.

'Our neighbours bake it,' Janina said. 'They give us a loaf that lasts us a week. My father does some tailoring for them in return. So we help each other.'

I finished my meal and thanked her very much. She said, 'It would be a big risk to stay at our place, but Marian will help you to organize your life in the forest. Later, after my parents go to bed, Marian will take you to the kitchen where you can have a wash.' I thanked her again.

After I washed, Marian handed me an axe, a short army spade, a hammer and some nails, which I put in my rucksack. He also gave me some potatoes, a bottle of water and a saucepan. I returned to the stable for the night.

Early in the morning I went to the forest on my own. I found my way easily and began to feel at home among the trees. I am going to make it, I thought. I collected some wood, cut it to one size, about a metre and a half, and started building a hut shaped like a tent. I covered it with branches and leaves. It took me a whole day to complete. I had to finish the job in a day because I needed a safe place to sleep. I did not want to go to Marian's stable every night, but that evening I had to return; I realised I didn't have anything to cover me or keep the entrance closed. The rabbits were at the stable as usual. Marian came in and sat down beside me.

'How are you getting on?'

'Fine. I finished my hut and have a place to sleep, but I need your old coats and a blanket. I also want three bricks to cook my meals on and a container to carry water from the river.'

Marian gave me the things I asked for. We agreed that from now on it would be better if I stayed in the forest and got food from friendly farmers in Siucice.

'Come to the kitchen,' said Marian. 'Janina will give you something to eat.' I had some soup, bread and meat, then I had a wash in a bowl of hot water. Janina gave me a towel to take with me to the forest. There was no sanitation in their house and water was brought from a well outside. When I got dressed, Marian asked me to come into the other room to see his parents. His father was a tailor. I got on best with the mother, who was well educated. She was in poor health.

'How are you finding it in the forest?' his mother asked.

'Fine: plenty of fresh air, wild animals and birds. I listen to the sounds they make. I feed the hares and deer with potato peelings.'

'We'll try and put some traps out for the hares,' said Marian. 'I'll prepare some wire and bring it to you.'

I was frightened to stay any longer in case they had visitors. I thanked them all and left. Marian came with me to the stable, checking in case anyone was around. He said, 'I'll come to the forest in a couple of days and bring you some more things, and some books to read. When you get to Siucice, be careful. Don't mention that you have been here.' I thanked him and said good night.

The next morning I went to my home in the forest. I took with me potatoes, onions and carrots, which I boiled for lunch. I made myself busy digging a hole in the ground, making a cellar to store my vegetables and masking it to make it invisible. I used the blanket to cover the entrance to my hut and protect me from animals.

In the afternoon I left my new home and went to my destination, Siucice. I took a short cut through the forest to avoid Blogie and joined the trail to Siucice. Walking along the road in the forest I heard horses and a cart coming behind me. When it got close I asked the farmer for a lift.

'I am going to Jaksonek, the next village.'

'That's OK,' I said.

'Get on. Where are you going?'

'Dabrowa, to my uncle to help him on the farm.'

'What is your uncle's name?'

'Zaborowski.'

'Is he the farmer who lives near the forest at the end of the village?'

'Yeah, that's him.' I knew the farmer, though he wasn't my uncle.

'I've known him a long time. Where do you come from?'

'Tomaszow.'

'Are the Germans there?'

'They're everywhere in the town.'

When we arrived at the crossing the farmer stopped the horses and I jumped off the cart. 'Thanks for the lift,' I said.

'God be with you.' He turned off into the village and I continued my journey. I passed by Dabrowa and another village and then had to cross a large common. Cows were grazing and two young boys were sitting at a fire. I greeted them and asked if I could join them. 'Yes,' they replied.

'You could bake some potatoes in that fire,' I said.

'That's what we are doing.'

I felt they did not trust me. I tried to be friendly, asking if they knew if anyone in the village needed a farm worker. 'No, we don't,' they said as they checked to see if their potatoes were ready. They passed one to me. I thanked them and told a story of how a friend and I were looking after cows grazing near the River Pilica. We caught a wild duckling and baked it in the fire with some potatoes. After catching it we killed it by cutting off its head, cleaned out the insides and covered it, still with its feathers, in clay, before cooking it. When it was ready to eat we took off the dry clay, the feathers came off with the clay, and we ate it.

I stayed with them and ate some potatoes until they were ready to head back to the village. I carried on on my way. I wanted to arrive in Siucice after dark so that no one would see me. I was still about five kilometres away. It was dark once I got there. I went to visit some friendly farmers who were surprised to see me, and curious as to how I had managed to hide. I told them I was staying in the forest so as not to endanger anyone and that I had to get food from villages, which I could only visit in the evenings.

During my visits to Siucice I stayed in different farmers' stables at night and in the day as I waited for the cover of darkness again. Late one night I was thirsty after a salty supper at a farm. I left the stable to go to the well. The farm dog came running towards me and I realised he was also thirsty. I went to the dog's kennel and filled its dish with water. After he had a drink I washed his dish and took it to the stable. I milked a cow into the dish and drank from it. I used dog dishes many times when I was hiding in villages. I was aware of the hygiene risks, but weighing it up, knew it was more important to get the nutrition from milk.

I always liked dogs and made friends with them when I lived in Siucice and worked for the farmers. This proved beneficial while I was hiding. Once I called them by their names they quietened down and stopped barking. I only took the dog dishes at night, aware I had to be very quiet, and only at farms where the dogs knew me.

Before I left Siucice I stole some linen from the gardens and doorways of three farms. I needed them as sheets to cover the moss and leaves that I slept on, and to cover myself. Farmers used to spin rough linen sheets from flax. They would dye them different colours and hang them over their doors in the summer to keep out flies, or use them as bed covers. I gave one sheet to Marian and asked him to arrange for his father to make me a pair of trousers. I had been wearing the same clothes for over a year and they were wearing out.

On the evening of the third day, I left Siucice with a rucksack loaded with food. I arrived home in the forest at about midnight. I was tired. After putting my rucksack in the storage hole I went to sleep in my hut. I fell asleep very quickly. Waking up early in the morning, I realised that I was lying in water. Only my head was above it, on a cushion made of leaves in a sack. I got out of the hut, took off all my clothes, squeezed out the water, and hung them out to dry on branches.

It warmed when the sun came up. I walked around naked all day. I had to remove everything from the hut and dry it. I raised the floor level above the ground by putting in more earth to prevent further flooding. After my arrival the previous night there had been a heavy downpour. As I was so tired I had not woken up. I was lucky not to have drowned.

That morning, nature was busily going about its business. Birds sang in harmony making a beautiful wild melody. I could hear the sudden roar of a deer, grunting boars and the short barks of hunting foxes. I wandered around, familiarising myself with the area. In among a bunch of saplings I found a lone baby deer. I sat down and played with it. I tried stroking it but it was scared. It was unable to run off; it could only have been a couple of days old. By sunset my clothes were dry. I got dressed and went to Marian's house.

All here was quiet. I went to the back of the house and peered through the window. I knocked on the glass and Janina let me in. Marian was out, but due home soon. She sat me down and gave me a drink.

'How's the forest? The rain was so heavy.'

I told her about the flooding. When Marian arrived I repeated the story.

'You must make something better,' he said.

'I've made some alterations, it won't happen again.'

'I have some wire for you for traps. I'll come to the forest tomorrow and show you how to make them. Did you manage to make some food?'

'Yes, baked potatoes. Tomorrow it'll be soup. I'll put some smoked meat in it. The food I brought with me from Siucice should last a couple of weeks. I need more potatoes, but I can take them from the storage mounds in the fields.'

Janina served up coffee, bread and butter and boiled eggs. She suggested I stay in the stable overnight and go back to the forest in the morning.

She woke me with a start telling me to run off. A policeman was at the house with Marian. I ran outside and walked behind the church fence, heading towards the forest. When I was some distance away I turned and saw the policeman and Marian standing at the back of the stable. Once in the forest I started running. I arrived at my hut and had to lie down and rest. I was upset; frightened that something might happen to Marian and his family. I could not relax. I was waiting for

Marian to come and tell me that he was okay.

To stop myself worrying I started to prepare food. I put water in a tin and washed some barley. When it was boiling I peeled potatoes, carrots and onions, washed and added them. After eating I collected dry wood. I looked for the baby deer, but it was gone. It had probably been taken away by its mother. I got back to my place about mid-day. After a while I heard the cracking of branches. I stopped and looked in the direction of the noise. I whistled to attract Marian's attention. We went to my hut and while we sat outside he took some wire from his bag.

'Would you like some barley soup? It's still hot.'

'Yes, please.' He wanted to know how I managed to avoid being spotted.

I wanted to know what the policeman was doing there.

'Nothing special. There were two of them on their way back from a patrol. One left to return to the station and the other hung around to have a word. Janina made him some coffee and then we drank some vodka. The policeman knows that I breed rabbits and asked if he could see them, as he wanted to breed some himself. We went to the stable and he noticed there was a place to sleep. I told him I sleep there if the Germans come to the village to avoid being taken off to a labour camp. He laughed and said that if something was up he would let me know.'

I took it as a warning that I should stay away from Marian's house for a while. I made another storage space and put in some of the food I had brought with me from Siucice. Marian was impressed with my work and how I masked it from view.

'Come, I'll show you how to lay traps for hares.' We went a few hundred metres into the trees, where he placed two traps. The traps were made from thin wire. One end was tied round the trunk of a thin tree the thickness of a man's wrist. Round the other end was a loosely knotted noose that dangled off facing branches growing out of the bases of two adjacent trees. Animals like hares and foxes used narrow paths worn into the forest floor and the noose was hung directly over one of these. The wire was covered with branches. Hares are nocturnal and run in the evening and night. When they ran into the trap the noose would tighten and strangle them.

'You have to check them every morning, that'll give you something to do,' said Marian. We went back to the hut and he left soon after. We decided that I would not go to his home for a while.

I had to go to the river to fetch water and go to the field for potatoes. I needed to make another storage area in the ground just for potatoes, which are best kept separate because they make other vegetables rot and smell. I kept carrots and beetroots and other foodstuffs together. I had to collect enough food that autumn to last all winter.

One day while I was just wandering around I heard the noise of rutting deer echoing through the forest. I crawled through the branches towards a clearing and could see several deer locking horns on a common while the females were grazing peacefully.

I started to think about making a place for the winter. The hut was going to be too cold, so I had to plan for something warmer. I lay down in my hut and drew up a plan in my mind. I fell asleep and when I woke, I realised the sun was going down.

I got dressed and went to the river to fetch water. When I got there it was almost dark, so I followed the river to the next village to stay overnight. I stopped at a farm to ask if they needed a worker. There was only a woman there.

'My husband is not back from Tomaszow. He went to the market. Would you like something to eat?' she asked.

'Yes,' I replied. 'I've walked all the way from Tomaszow. I couldn't find any work, or food.'

I ate. We could hear the farmer as he arrived soon after. A young boy came in first. His mother greeted him. He told her what they had bought and what they had sold. Then the farmer came in. His wife introduced us and explained that I was looking for work.

'I don't need anyone at the moment. Later in the summer, maybe.' He said he didn't know anyone else who did either.

He told his wife that he had heard that the Germans and the police were raiding villages along the river for farm produce. I was suspicious about all this talk of Germans, believing it was intended to encourage me to leave. I stayed a little while longer, thanked them for the food and left.

On the way back I spotted a milk can hung on the fence of another farm. I took it and walked along the river to try and find my hidden water tin, but it was too dark. I decided to stay the night on a raft by the riverbank. It was a very long raft, used for transporting trees along the river. There were several huts on it. I went into one and found plenty of straw to cover me.

It was very cold, and I could not fall asleep. I waited until it was light and then looked for my tin in the forest. I filled it with five litres of water and put 20 litres in the milk can. I carried it all back to my home. I was hungry, so I made a fire and put potatoes in to bake. Then I boiled some water for a hot drink.

The days went by with a routine of eating three times a day, doing exercises and working on my new storage bunker. Every morning I checked the traps, but they were empty. I put potato peelings in different places near the traps. One morning I went to check them and could not believe my eyes. There was a hare with all its insides missing. Marian explained that a fox would always get to the trapped hare first and eat its tastier insides. They'd come back for the meat when it was rotting. I took the remains of the hare to my place, skinned it, washed it, salted it and left it for some hours. I knew I could be at risk from rabies, which was why I salted it. Then I made a fire and cooked it, turning it on sticks. At last I had a proper meal. I kept some aside and put it away in my new bunker. I'd catch a hare once or twice a month.

Going Underground

One night in June 1942, while I was asleep, my hut fell in on me. I woke up terrified and to my surprise heard wild boars and piglets running away. I did not sleep the rest of the night. At daybreak I started to rebuild my hut. I had to demolish what was remaining and start afresh using the same tree trunks. I kept the tent-like shape but rooted the wood deeper into the earth. It arose in a slope on each side a metre above my head and this time I secured the wood to a central beam with nails. I covered it with moss and branches to keep it hidden and waterproof. Inside, there was just enough room for me to lie down.

I worked on it until lunchtime without bothering with breakfast. When the structure was finished I prepared something to eat. I cut potatoes into small pieces to boil them more quickly. When they were soft with only a little water left I put some flour on top and boiled it all a little more. I mashed everything up with a thick stick of wood. I'd seen something similar being made in the villages, though they put bacon and fat on top. I could only add a little salt, but it was still a tasty meal. I felt satisfied and capable of carrying on.

After eating, I lay down to sleep but was unable to drift off. My head was full of plans on how I would build an underground bunker. It was a short nap; I was compelled to get up and choose the best spot to build my bunker. I found somewhere not far from the hut, positioned between two rows of trees so that I did not need to cut any down and would be well hidden. As I was familiar with this part of the forest I knew I'd be able to find my way home.

I started straight away by removing moss from the forest floor and marking out where I needed to dig. I loaded earth into my rucksack and carried it deeper into the forest where I found a dip in the ground. There I brushed away the moss and leaves and emptied the contents of the rucksack. I repeated this many times until the sun started to come down. After a rest I boiled some water for a hot drink and ate a chunk of bread.

In the twilight I made my way to Marian's home. It was dark when I arrived. Janina made me some barley coffee and boiled eggs with bread and butter. I sat down with Marian in the kitchen and explained why I had decided to build a bunker. I drew him my plans on a sheet of paper. He approved the idea and design and told me to make sure that two people would be able to sleep there in case he needed to run away from the village. I told him I had taken account of this in my original plan. I had worked out everything in fine detail, including what I needed for cooking, washing, keeping warm and ventilation.

'Tell me Marian, where can I get a metal plate to put on my fireplace?'

'I can tell you where to find a broken metal plate with rings.'

'Have you got some pieces of pipe I can use for ventilation and an outside chimney?'

'No I haven't, but you can find pieces of pipes from old church organs round the back yard of the vicarage.'

'How do I get there?'

'You have to climb over the back fence in the evening and walk along the garden to the corner behind the barn and there you'll find everything you need.'

I left Marian soon after and went back to my hut in the forest. It was as I had left it. I climbed in, covered the entrance, and went straight to sleep. The next morning I got up, made some breakfast and, after that, I started to dig and continued carrying away the earth. Half a metre down below the soil was hard clay that I had to chop with my axe. I worked like that for a week to remove sufficient earth and clay, which I finally covered with the same moss that had originally been there. To make sure animals would not come and dig up the earth, I covered it with broken trees and branches.

When I finished the digging the sleeping area was about one metre deep and over a metre wide, enough for two people to lie side by side. The bunker was not high enough to stand up in; I could only sit, lie or crouch.

I went to other parts of the forest and cut down thin trees around 25 metres in height. I cut batons to measure, about half a metre each, from the tops of the trees, managing three or four batons per tree before the trunk got too thick. I carried the batons to my bunker and stuck them about five centimetres into the clay on the floor of the bunker and about five centimetres apart. They were to insulate the site. I left a ten-centimetre gap between the batons and the clay and about five centimetres between each baton. The gaps were filled with dry moss and leaves. The whole bunker needed about fifty batons.

On the floor of my sleeping area I lay the thinnest batons tightly together. I made the floor a little lower in the middle to give the batons some spring.

Then I built a triangular roof frame from sloping batons. The main support beam was a tree trunk about three metres long and more batons were placed with small gaps along the frame. The beam rested where the sloping batons met and was nailed in. The sloping batons went over the vertical side batons and into the base of the bunker walls. The roof batons were covered with moss and leaves and I made putty from clay and water, which I still had to carry in a tin and a 25-litre can from the river, five kilometres away, through the forest. I covered the entire roof surface with clay to make it weather-proof and smoothed it all out to make sure any moisture would slide off.

The support beam ran ten or fifteen centimetres beneath the forest floor. I filled in the area above the sloping roof and over the beam with earth, clay and leaves to make it level with the ground so that anyone who walked over would not detect what lay underneath.

I built an entrance to the bunker and cut out steps leading down to it in the clay. The bunker was shaped like a cross. The steps were at one end. They led into the sitting and cooking area, the arms of the cross, and from there you went into the sleeping area, which I had to slide into head first as if I was entering a baker's oven. Over my head were two thin metal pipes from the church organ that I used for ventilation when

I closed the entrance cover. The pipes were at a 45-degree angle on either side of my head, piercing the ground just away from the bunker.

The entrance cover was a half-metre square wooden frame doubled up with moss and leaves. It was in the ground like a trap door and had a piece of string attached to lift one end. I would put it to one side when I entered and, at night, closed it from the inside like a loft flap.

The sitting area was dug out of the clay and insulated for my back with wood, moss and leaves. The floor of the sitting area was about half a metre deeper than the sleeping area so my legs could dangle comfortably. At head level was a shelf where I kept dried barley, beans and flour, etc.

Opposite the seated area was a fireplace cut out of the clay and covered with a broken half plate and rings to cook on. A chimney about ten centimetres square, made out of little stones I had collected with difficulty from a wide area in the forest and bound together with clay, rose from behind the fire to ground level. When the fire was burning I connected the pipe outside to take the smoke higher, about 30 centimetres above the ground.

When I left the bunker, I pulled the chimney pipe out of its clay socket and kept it in the bunker, covering the chimney hole with wood and spreading leaves on top. I also covered the entrance with leaves. I covered over the top of the bunker with broken branches and leaves to make it look natural, and difficult to walk over. In all, it had taken me about two weeks to complete the bunker during July and August 1942.

I covered the batons in my sleeping area with dry leaves and moss for a mattress and laid a linen sheet over it. I covered myself with another sheet for sleeping. I usually slept in my clothes and was always warm enough from the fire. I felt safe from the first night I slept there and was satisfied with my handiwork.

One day after sunset, I went to visit Marian and told him I had completed my task. He was very keen to come and view my new home. Soon after he came to visit, and he was amazed at how well I had planned and completed the work. He wondered how he'd find me if he needed to come into the forest while I was asleep. I said I would leave a water can outside as a marker.

Throughout that summer, I visited Marian regularly and went to Siucice to get most of my food. During my journeys through the villages I collected other things I needed for my home. One day, in a neighbouring forest not far away, I noticed a hole in the ground. It was about a metre in circumference and the earth had fallen in. In one corner I could see slats of wood at right angles. It looked like it had once been a well. I asked Marian about it. He explained that where my forest was, there had been a village before the First World War, when half of Poland had been occupied by the Russians and the other half by the Austrians: this had been demolished by the Russians, who then planted the forest in its place. There could indeed be an old well.

On the following day, I started to investigate and found clean sand below the earth and moss. To my surprise, the hole was full of water and teeming with little green frogs. After killing the frogs I scooped out all the water down to the clear sand. I came back in two days to find the hole full of clean water and free of frogs. I tasted it; it was good. I no longer needed to fetch water from the River Pilica.

I caught another hare that I took as a present to Marian. When he came to the forest and saw my home, he was very impressed.

Sometimes, when Marian found out from the police that the Germans were due in the village the following day, he would go into the forest under the cover of dusk and come and spend the night with me in the bunker. It was dark by the time he'd get to me and on several occasions, when I was already asleep, he walked around the forest for maybe two or three hours, whistling as he tried to find me. I'd wake eventually and, recognising his whistle, would light a thin piece of wood, open up the cover of my bunker and hold it up for him to soon spot. He'd complain, shouting, 'I've been walking for hours and couldn't find you. You're just like a wild animal buried away.' I pretended to be sorry but I wasn't. I could find my bunker with my eyes closed.

'You know Maniek, you could make another bunker, a bigger one with more places to sleep. We would then have a spare place in another part of the forest.'

I agreed, and started to build a second bunker, which I finished before winter. I was working hard and felt that I needed more meat to eat.

The previous winter, when I was hiding in Siucice, I had stayed for two weeks with Farmer Piatkowski. I slept in the stable at night and thrashed his grain during the day. When I finished he told me to go away. I asked him for some money, but he said, 'Why? We fed you, my wife washed your shirts, that's enough.' I said that I needed to buy a toothbrush and toothpaste, but he gave me nothing.

Now, six months later, I returned one night to Piatkowski's farm in Zawada, near Siucice. I lay down in his stable and slept until the farmer came in. I woke up. Piatkowski was surprised to see me and I could see he was not very happy.

'I will not stay long at your place,' I said. 'I am now staying in the forest, not in the villages. I need some food for me and others.' I was implying I was now with the partisans and he immediately changed his attitude saying, 'Stay the day at the farm and in the evening I will give you some food and you can go.' He was worried that if I left during daylight other farmers would see me. He gave me some smoked meat, dry beans and barley.

His wife asked, 'Would you like some flour?'

'Yes please, it would be very useful.' In the evening, they gave me some food and I went on my way to my forest home.

I found some mushrooms and took them to Marian's. He told me that I had to boil some, then throw away the water, add fresh water and make a soup with it. I went on collecting mushrooms and made soups. One day, after eating one batch of soup, I fell ill, and realised that I had been poisoned by bad mushrooms. I lay in my bunker and drank only water for two days. I survived, but was unable to leave the forest for several days. It occurred to me that if I died, I was already in my grave, so at least nobody would have to bury me.

Marian came to visit me, worried at my absence, and I told him what had happened. 'You have been lucky to get over it,' he said. 'You must be very careful. Probably better not to eat mushrooms for a while.'

I started baking cake from flour and water on my hotplate

every morning and evening and boiling my mid-day soup. I was soon fully recovered.

Marian said, 'You need to prepare yourself for the winter. Once the snow has fallen you will not be able to leave the forest because you will leave tracks and somebody could find your hiding place. You must collect enough food to last you from November to March.'

'So that means I'm going to have to be a full-time hermit?'

'You have no choice,' said Marian.

I made myself busy throughout August collecting dry wood, chopping up fallen and dead trees. I kept to my old habits, cutting everything to the same size and stacking it neatly between two trees. The top was sloping for the rain to run off and I covered it with leaves and branches.

In September, I began collecting food for the winter. On my way to Marian from the forest I walked along some fields and eyed what was growing. Going back in the evening and early morning I collected vegetables in my rucksack and carried them to the forest. I put them into my cellars: potatoes in one and vegetables in the other.

Sometimes I did not make it to Marian's. I got to the edge of the forest and, hidden from view, watched the farmers working in the fields. When they had gone home, I moved in and dug up vegetables to take back to the forest.

Day by day, week by week, I went on with the same routine: preparing food, eating it, reading books, going for long walks to exercise my legs, familiarizing myself with a large area of the forest, and working on the second bunker. I observed the behaviour of wildlife, the birds building nests and feeding their young. Some days I met the same old male boar. Adult boars were black, but he was so old he was grey. He did not run away, he just walked very slowly. Marian told me this old boar was respected for his age by the forest rangers who warned people not to shoot him. He did not appear scared of me but I was wary and kept my distance. One day I met a female boar with about ten piglets, they were brown with dark stripes. I had to run away and jump up a tree as she ran towards me. She got to the tree and grunted. Her piglets came after her and a few minutes later she went off, allowing me to clamber down.

Marian taught me so much about how to survive in the forest, imparting some useful bit of information every time we met. I had not been to Marian's home for a while. One afternoon I was collecting firewood when I heard somebody walking towards my bunker. I lay myself down on the ground and saw it was Marian. I went to meet him and together we walked to my place.

I asked him about the news and how the family was. Marian said his mother had died peacefully during the night six days previously. He asked me to make a cross from white birch wood and bring it to his home the next evening. He wanted it for her grave. He told me that most days his mother had been asking after me, wanting to know how I would survive in the forest alone.

I showed Marian the stock of food I was preparing to put in my cellars for the winter, and the firewood I had covered over and camouflaged between the trees. 'How is your work on the second bunker getting on?' he asked.

'Come and see for yourself.' We went to a different section of forest about 300 metres from my bunker where I had dug the hole for the new bunker. It was three metres square, about three times the size of my one, and much deeper so that we could walk inside. In one corner I planned to build a fireplace. This time I had to cut down some trees.

Marian approved of my work. I hoped to have it ready before winter set in. It would be impossible to work on it once the snow had fallen. The most difficult part was carrying away the earth and clay and hiding it so as not to leave any marks.

We went back to my place and I boiled some water for the coffee Marian had brought with him. He left before dark. I accompanied him to the edge of the forest and looked around to check no one was on the road. It was safe. He left, saying, 'I will see you tomorrow evening.'

'Yes,' I replied, 'I hope so.'

After Marian left I went back to my bunker and started to make supper. I baked a cake from flour and water on the hot plate and boiled some water. That was my usual bunker supper. When it got very dark I started to read a book by the firelight. There was no other light. Before I went to sleep I closed the entrance and had the air circulate over my head. It felt as though I was sleeping in a grave. I was thinking and planning my work on the other bunker and thought that Marian must have a good reason for asking me to build it. I knew he was in contact with the underground and maybe he was planning to organize something. He told me during one conversation that, in some forests, partisan groups were already organizing and operating.

Next morning, after breakfast, I went out to cut a thin birch tree for a cross before continuing work on the second bunker. I made a professional job of the cross, cutting into the wood so that I did not need any nails. That evening, I took the finished cross to Marian. I arrived at his back garden as usual. Marian came out and appeared satisfied with the cross. He put it in the shed and invited me into the kitchen. Janina was there and seeing me, drew the curtains and prepared me some food. I expressed my sympathy over the death of her mother.

Marian told Janina that I had made a cross and had brought it from the forest. She thanked me for my effort. I told her that I had more reason to thank her for all that she and Marian were doing for me.

That evening, I went back to my place in the forest. I told Marian that the next day I would go once again to Siucice and collect more food for the winter.

Janina wondered how I got on so well with Marian. I was always respectful towards him. He'd done his National Service and had become a sub-lieutenant, and he liked the way I deferred to him like you would a superior officer. In return, he respected my abilities as a handyman.

I knew I had to keep him calm and didn't do anything that would upset him. If I knew I was right about something I made sure I said it diplomatically, that way he could accept what I was saying. I never said he was wrong because that would make him angry.

He was fascinated by gliders and asked if I thought I was capable of building one. He had the idea that we might be able to fly through the night, high in the sky to England. I explained that it might be technically possible to build it if we could get the right instructions and could find a large enough workshop. I didn't really think it was possible, I was just humouring him.

I managed to convince him that the biggest problem would be taking off. We didn't have anyone capable of pulling it far or fast enough, or a large enough common to give it enough of a clearing. Eventually he gave up the idea. He'd mentioned it to a friend in the underground who convinced him it was impractical.

A New Identity

It was near the end of September, the nights were getting colder and the forest floor was thick with fallen leaves. I made my way to Siucice and decided to stop over at Stefan's. His mother, Szyszkowska, who owned the farm, was a widow. I had slept there many times when I first hid in the area; they were very good to me. They tied their dog on a chain with a ring through a wire so he could run along the yard. Some evenings Stefan's mother took the risk of inviting me into the kitchen. The farm was away from the village and the road and if anybody approached, the dog would bark and not let them pass. That gave me time to hide.

Stefan was about thirty years old, slim and tall like his mother. Sometimes we'd sit all night distilling vodka made from rye grown on the farm. I would watch as he mixed the rye flour with water and yeast. It would be left to rise for about a week. By then you could smell the alcohol as it fermented. The mixture was put in a sealed kettle with a pipe coming out, heated and the steam passed via a copper spiral pipe through a barrel of cold water. At the end of the pipe the alcohol dripped into a jug. The alcohol was put back into the cleaned out kettle and the process repeated leaving 70 per cent proof vodka. This was then filtered through charcoal to get rid of all the smells. We tested the vodka as it went through the process, eating boiled carrots as appetizers. The carrots were boiled on top of potatoes prepared in a big saucepan to feed the pigs and cows. The rye mush left over from the vodka making process, free of alcohol, was mixed with the mashed potatoes and oats for the animals.

Stefan had recently married Stasia, who I'd known for some time. I had met Stefan sometimes when I visited her farm where she lived with her sister and widowed mother in the nearby village of Zawada. They had let me stay at their farm. If Stefan had married a girl I did not know I could not have risked staying at his farm any more. Any hiding place was important to me. I had to entrust them with my life.

I arrived from the forest in the evening and the dog started barking as usual. I called out his name, 'Burek'. He grew quiet and jumped on me as a friend. Stefan heard the barking and came out. I asked him if any outsiders were about. There was only family. He brought me into the kitchen where I was greeted warmly.

'How are you managing, Maniek?' asked Stefan's mother.

'Not bad,' I said. 'You know I had to move away from here because too many people knew that I was hiding in this area and Sieczko was after me. I'm hiding in the forest. I wondered if you could help me with some food?'

'We will give you what we can and you take care of yourself,' said the mother.

I stayed the night, had some food, washed and went to sleep in the stable. Stefan told me to be very careful. He heard that the shoemaker from Siucice had been hiding at a farmhouse in another village with his wife and two boys. The farmer put them in his cellar and took all their jewellery and money in payment. One evening, the farmer took them out to the fields for a night and a day, saying he had heard that Germans were coming to the village. He left them there to sleep and after a short break came back with an axe and killed the wife and one boy. The shoemaker managed to escape with the other boy. They ran to Siucice and told the farmers there what had happened. The shoemaker went mad and became terrified of everybody. The shoemaker and his son stayed mostly in the forest during the day, returning to the village in the evenings to ask for food and a place to sleep.

One morning the Germans came to the village, and when the shoemaker saw them he ran through the fields towards the forest. The Germans started shooting and killed the father and son. This news upset me very much. I knew that the two Jewish brothers from Skorkowice were hiding in the neighbourhood. I had met one of them, Jankel, while we were both on the run. Stefan did not know where they were.

Stefan also told me that at Treblinka, transports were arriving every day full of Jews and they were being cremated. It was final confirmation of what I already knew in my heart; that my mother and sister must be long dead. He did his best to be sympathetic.

Next day, Stefan's mother gave me dry beans, barley and some smoked meat. I waited until the evening then made my way to Siucice. I went to farmer Lesniewski, who had a flour business. I asked him if I could get some flour. He said 'How much can you carry?' 'About two kilograms would be enough for me.'

He gave me some flour in a cotton sack while his wife prepared some food and drink. I thanked them and went out into the dark. I knew them well as I had done some work for them on the farm when I used to live in the village. They liked me, and had always helped me, but I could not stay at their farm.

The last stop I made was at Morawska's farm, which was a good departure point for joining the route to Blogie. I stayed overnight and through the next day. At sunset, with a rucksack of food, I started my journey back to my forest.

Late at night, not far from Blogie, I heard the squeaking of a bicycle coming towards me. It suddenly stopped. A voice called out, 'Stop! Police! Who's there?' Making my voice deep, I replied, 'Stay away. A partisan group is on its way and we will shoot you if you get any closer.'

The man on the bicycle turned around and went back the way he had come. I could not believe that I had been so brave. I was shaking, but elated at my success. I wished to be a fighter and take revenge for all our people who had been killed.

A short way on, I turned into the fields and the path leading to the forest. I arrived at my place after midnight. I found my bunker, entered it and lay down. As I was very tired, I should have fallen asleep straight away but I could not. I was still thinking about my encounter with the policeman.

I was worried about being able to survive over the winter in the forest without any contact with another human being; but I clearly remembered my mother's words – as we were lined up by the Germans – when I was unsure what to do. 'Go my son. Step out.' And my sister added, 'Maybe one of our branches shall survive.' I wanted to survive to fight the occupiers; the murderers of maybe millions of innocent people.

Eventually, I fell asleep in my living grave. I woke up later that morning tired and hungry. I went for fresh water and breakfasted on the bread that I had brought from my journey. I sorted out my storage space and rested for the remainder of the day. The following days I worked on the other bunker and prepared wood for the winter.

By late October I had nearly completed the second bunker. When I finished the cover and had masked over the top I went inside to check things, walking around, satisfied with my handiwork. Coming out, I stood in the entrance, which still needed a flap. As I looked around there was a sudden crashing noise behind me; I was horrified to see that the bunker had caved in. If I had been inside I would have been buried in a grave of my own making.

I did not start to rebuild it straight away. I returned to my first bunker, prepared some food and laid down to rest and think over what I had done wrong. In the evening I went to Marian's home. I told him what had happened at the bunker. 'What did you use for support in the middle?' he asked.

'I cut down a dead tree which was dry. I didn't want to destroy a living tree.'

'That was your mistake,' he said. 'You should have taken a living tree to make a safe post in the middle of the bunker.'

I stayed for a short time, had some food and went back to my bunker. The next day, I started to rebuild the cover of my new bunker. I had to take off all of the top pieces and cut a stronger tree to rebuild it. The most difficult part was to replace the clay that covered the top, and to make it safe from water leaks. It took me several days to complete, interrupted by my early evening forays into the fields to collect potatoes.

After some days of checking my new bunker to ensure that it was good and safe, I masked the top with branches to make it impossible to walk over. I started to insulate the walls inside and make a cover for the entrance. I built a chimney and made a fire. Everything was in working order.

The nights were getting colder and the days shorter. I decided to go to Marian's house and got up earlier than usual to collect mushrooms, taking care the ones I picked were safe and not poisonous. By sunset I began my journey, laden with mushrooms.

Janina was very happy about my delivery. I had something to eat in the kitchen and told Marian that I had finished rebuilding the second bunker. He said he would come the next day to check it out. I asked him to bring with him various things for storage, matches, salt etc., before I made my way home.

The next lunchtime, Marian arrived in the forest and brought what I had asked for. I made lunch and then we went to the other bunker. Marian was very impressed. I asked him how long the bunker would stay empty. 'I hope not for long,' he replied. 'I have a plan. I will explain everything later.'

I told Marian about my encounter with the policeman on my way from Siucice. He laughed and said, 'I'm not surprised. The police know there are some partisan groups and they are very frightened. They have very few weapons. They would not like to fight Polish partisans, being Polish themselves. They have to do their job as they are forced to by the Germans.'

'I would rather fight the Germans than stay all winter alone in that bunker.'

'We can't do anything now, but I have something in mind,' said Marian. 'Would you really like to fight the Germans?'

'Yes,' I assured him.

'First you must choose a name and forget your Yiddish language.'

'How?' I asked.

'You must talk to yourself in Polish and read books in Polish. What name will you choose for yourself?'

'As I was born in May, I'll be Mayevski,' I announced.

Marian grinned. 'And which day were you born?'

'The fourth of May.'

'That's St Florian's Day,' said Marian. 'The patron saint of the fire brigade! So from this day forward your name will be Florian Mayevski.'

'Yes,' I said as my new name sank in. 'I am Florian Mayevski.' I tried to forget I was Moshe Aaron Lajbcygier. Unlike most Jews who spoke Yiddish as their first language, I never spoke Polish with an accent. My Polish language was very good; I'd always mixed with Polish boys.

'You have to come to visit us so that I can introduce you to Janina with your new name.'

We discussed the practicalities of my staying in the forest through the winter. I would not be able to go for water. I would have to collect snow and melt it over the hotplate and then sieve it through a linen cloth that Marian had brought for me.

Before dusk, Marian left for home. I went back into the bunker, sat down in my sitting place and made a fire. Sitting and watching the fire as it got darker outside, my whole precarious predicament suddenly came home to me. I started to cry like a baby, sobbing to myself, 'Mother, why did you let me stay alone? I miss you so much. God help me survive. I want to know what has happened to you, Mother, my sister and all our family.'

I stopped crying, and while watching my fire burning I thought, 'I must be strong. I will not give up. I will fight for my life and my freedom.'

I came back to reality and started to prepare my supper. After eating it, I read a book until I was tired. I carried out my ablutions and went to sleep.

Next day I decided to go to Marian's home. It was the middle of November and there was not much time before the onset of winter. Janina gave me food and drink. She congratulated me on my new name and the job I had done in the forest. Marian had told her all about it.

Winter Hibernation

I went back to the forest and continued with my routine until the middle of November when, opening my cover first thing, I was taken aback by a splatter of snow falling on my face. Halfway up the bunker step I looked around with heavy eyes. It was not cold, in fact it was rather mild. The air seemed cleaner and the atmosphere subdued. Only the birds conversed.

The forest floor was so dazzling I had to look away. The trees were weighed down by their winter blanket. It was so beautiful, like being inside a giant painting. I was quite shocked.

I was now a full-time hermit. It would be too dangerous to leave the camp; my tracks could lead someone straight to me. The nearest track was just 150 metres away, though it was used rarely. I could hear farmers coming from afar on their horses and their carts with their metal-hooped wheels. There was no chance of them detecting me through the dense bracken and trees, but I couldn't risk them spotting my footprints.

I could venture no more than ten metres from my bunker and had no need to. The two full wood stores were on the perimeter and the two food cellars were just two or three metres from the bunker.

More snow fell at night and started to freeze. I had to keep my fire going until it was time to sleep. The bunker could become very warm, but once the fire was out, I often had to block a ventilator to stop cold air blowing in.

Mostly I washed by rubbing snow all over my body, or sometimes in warmed melted snow. I boiled my clothes in a saucepan that was not used for cooking. I made improvised soap flakes by scraping home made bars of soap I got from the village.

When I used up all my stored water from the well, I melted snow and filtered it through linen to purify it. The snow looked pure and white, but it was polluted with the soot of war, leaving the linen black.

Every day I would do my exercises to keep fit. I got used to the new routine, washing, baking unleavened bread from flour and water, and cooking meals. When I finished my smoked meat I started to look for hare tracks in the snow and putting down traps.

I could hear wild boar ploughing up the snow looking for food. I marked out my territory by walking around. I wanted the animals to be frightened by the sight of human tracks. I had to keep doing it whenever more snow fell.

I lost count of the days; I did not know when it was a holiday or even New Year. Some days I overslept. I could sleep for 24 hours. My body was slowing down; I began hibernating like an animal. I wished I could sleep the whole winter through, as a bear does. But I had no choice, I still had to eat and exercise. I knew it was time to rise when I saw light through my vent.

It was important for my sanity to keep to my routine. I kept wishing the snow would melt so I could end my isolation and visit Marian and Janina in Blogie. I could not stop thinking about my family, the War and my predicament. I had conversations with myself, talking out loud, discussing how I could have saved my brother by not letting him go to Sulejow, how I could have saved my mother and sister. I berated myself for not having done enough, even though I knew there was nothing more I could have done.

I was frightened of going mad through not saying anything to anybody else for all those months. When such thoughts came to me I tried to expel them and come back to reality. Reading helped a lot, calming me and keeping my mind off other things. I knew if I allowed myself to think too much I would get upset.

I became like an animal, my eyes and my ears always watching and listening for something unusual. I could hear sharper and more distant sounds; I became aware of the